3/ 2015

22/03

3/16/15

(32)

THE MEXICAN AMERICAN
FAMILY
ALBUM

THE MEXICAN AMERICAN
FAMILY
ALBUM

DOROTHY AND THOMAS HOOBLER

Introduction by Henry G. Cisneros

OXFORD UNIVERSITY PRESS • NEW YORK • OXFORD

Authors' Note

In this book, we have used the term *Mexican Americans* to refer to residents of the United States whose ancestors were Mexican citizens. Mexican Americans themselves have used a variety of names to indicate their heritage. In the 19th century, those living in California called themselves Californios; those in Texas, Tejanos; and those in the area that is now New Mexico and Arizona, Hispanos. These words appear as they were written in some of the selections in this book, and for clarity we have used them in introductions as well.

In the 1960s, many Mexican Americans adopted the term *Chicano*. The origins of this word are somewhat unclear. Some authorities believe that in Mexico it was a slang word for the very poor. As used in the United States, *Chicano* became an assertion of ethnic pride. Not all Mexican Americans, however, favored its use.

Today, the term *Latino* has come into general use, but it usually refers to a broader group than the Mexican Americans alone.

Oxford University Press

Oxford   New York
Athens   Auckland   Bangkok   Bombay
Calcutta   Cape Town   Dar es Salaam   Delhi
Florence   Hong Kong   Istanbul   Karachi
Kuala Lumpur   Madras   Madrid   Melbourne
Mexico City   Nairobi   Paris   Singapore
Taipei   Tokyo   Toronto

and associated companies in
Berlin   Ibadan

Design: Sandy Kaufman
Layout: Greg Wozney
Consultant: George Sánchez, associate professor of history and American culture,
        University of Michigan; author of *Becoming Mexican American*

Published by Oxford University Press, Inc.
200 Madison Avenue, New York, New York 10016

Oxford is a registered trademark of Oxford University Press

Library of Congress Cataloging-in-Publication Data

Hoobler, Dorothy.
        The Mexican American family album / Dorothy and Thomas Hoobler
        Introduction by Henry G. Cisneros
        p. cm. — (American family albums)
        Includes bibliographical references and index.
        1. Mexican Americans—History—Juvenile literature. 2. Mexican American families—Juvenile literature
        [1. Mexican Americans—History] I. Hoobler, Thomas. II. Title. III. Series.
E184.M5H66   1994   94-7785
973'.046872—dc20   CIP
                                                                                AC
ISBN 0-19-508129-3 (lib. ed.); ISBN 0-19-509459-X (trade ed.); ISBN 0-19-509125-6 (series, lib. ed.)

9 8 7 6 5 4 3 2 1
Printed in the United States of America
on acid-free paper

Cover: The Ramirez family of Nietos, California, around 1890.

Frontispiece: A 19th-century Mexican American family in California at the christening of the youngest child.

Contents page: Raquel Echeveria, Los Angeles, 1920s.

CONTENTS

Introduction by Henry G. Cisneros _____ 6

CHAPTER ONE
THE FIRST MEXICAN
AMERICANS _____ 9
The Mexican North _____ 12
Strangers in Their Own Land _____ 16

CHAPTER TWO
GOING NORTH _____ 23
Life in Mexico _____ 26
The Decision to Leave _____ 32

CHAPTER THREE
BACK AND FORTH _____ 39
Coyotes _____ 42
Evading the Migra _____ 44
A Foot in Two Countries _____ 46

CHAPTER FOUR
GOING TO WORK _____ 53
Finding Work _____ 54
On the Rancho _____ 56
Railroads and Mines _____ 58
Food for the Table _____ 60
Prejudice _____ 66

CHAPTER FIVE
PUTTING DOWN ROOTS _____ 71
El Barrio _____ 74
La Familia _____ 78
School _____ 84
Religion _____ 86
Associations _____ 90
Depression and War _____ 92

CHAPTER SIX
PART OF THE UNITED STATES _____ 97
La Huelga _____ 100
The Chicano Movement _____ 104
Newcomers _____ 108
Making It _____ 110
Celebrating the Heritage _____ 114
The Acosta Family _____ 118

Mexican American Timeline _____ 122
Further Reading _____ 123
Index _____ 126

INTRODUCTION

by Henry G. Cisneros

High in the Pyrenees in the rebellious Basque country lies a town called Munguia. It is said that sometime in the middle 1500s two brothers left their home, braving the vast Atlantic to seek their future in the promising new world.

After reaching the capital city of the Aztecs, what is now called Mexico City, one brother continued west to *La Perla del Occidente,* Guadalajara. The other went north along the trails that would eventually lead to cities in northern Mexico and what is now the United States. These brothers were, I am told, the forebears of my maternal grandparents.

The origins of my father's family are a little harder to establish, but we think that they may have been among the hardy travelers who crossed the frozen Bering Strait, always traveling south, seeking warmer climes along the blue Pacific. They reached the deserts of Baja California, turned east, crossed the purple majesty of the Rockies, and reached the lush lands of today's New Mexico. There, over centuries, they intermarried with the Spanish colonists and helped create a new mestizo culture.

These are the stories I tell my children. They may be partly fanciful legends and myths, but at their core is truth. Children should know where they come from, so that their pride and dignity will anchor them in a world where change and modernity can disorient us.

My paternal grandparents, Pablo and Crecencia Sisneros, na-tives of New Mexico, went north to Colorado, where they raised their family of 12. Only one child went to high school and found his future away from the fields, the herds, and the flocks that his family lovingly cared for. That was my father, George Cisneros. He won a scholarship to a Denver business college and entered the federal civil service. He joined the army in 1941 and after wartime service in the Pacific, he was assigned to Randolph Field in San Antonio. There he met my mother, Elvira Munguia.

My maternal grandparents, Romulo and Carolina Munguia, were a unique pair. Romulo was a staunch liberal, a revolutionary who helped overthrow the Díaz regime in 1911. Carolina, educated at a Methodist college in Puebla, Mexico, was the cherished daughter of a prominent family.

Romulo Munguia was apprenticed to a publishing house in Mexico City, where he learned the typesetting trade that he practiced for 70 years. The Mexican government sent him to the Mergenthaler Linotype factory in New York to learn the new linotype machine, which could set type far faster than it could be set by hand.

Since he knew no English and his instructors knew no Spanish, he was given a linotype and a box of tools and told to take it apart and re-assemble it, again and again. Until he was 90 years old he could do it blindfolded. He took this valuable knowledge back to Mexico.

As demands for democracy in Mexico grew, as people clamored for liberty, Romulo was in the fore-front of the labor organizing movement. He helped organize the first printers' and linotype operators' union in Mexico City as well as tailors, seamstresses, restaurant workers, and transit and electrical workers. The movement spread like wildfire and formed the grassroots support that brought about the abdication of Díaz and the establishment of democracy.

Romulo Munguia participated in the first Constitutional Congress in 1917. His revolutionary activity took him throughout southern Mexico. He helped local governments publish newspapers, organize meetings, educate the indigenous population, and supervise the massive programs of welfare, food distribution, and housing needed in the war-ravaged communities.

Romulo left government service in 1920 and devoted himself to his own newspaper publishing business, but his opposition to the Obregon and Calles regimes caused him to flee to the United States in 1925. He found a job with *La Prensa,* the leading Spanish-language newspaper in San Antonio, and my grandmother joined him the next year. My mother was two years old when the family settled in West San Antonio, in what was known as *la colonia Mexicana.*

I feel certain that at first my grandparents suffered from their inability to speak English, but they never, to my knowledge, complained of discrimination. When they encountered obstacles, they took them as challenges and worked to overcome them. They organized study groups, and they supported their schools and

Secretary Cisneros with his parents, George and Elvira Cisneros.

churches. Carolina organized Spanish-speaking PTAs in San Antonio schools where language proved a barrier to Mexican American participation. To assist the needy and combat discriminatory practices, Carolina and other PTA mothers worked with Mexican American merchants and barbers and their Anglo friends to provide free lunches, haircuts, showers, and clothing—well before Franklin Roosevelt's New Deal programs of the 1930s.

In their later years, Romulo and Carolina devoted much of their energy to improving relations between their two countries, Mexico and the United States. Romulo initiated and for 20 years organized annual seminars by professors of the National Autonomous University of Mexico (UNAM). His labors were rewarded by the establishment of a permanent branch of UNAM at San Antonio, where today several hundred students study the language, customs, culture, and history of Mexico.

I enjoy telling of the contributions and triumphs of these Mexican immigrants—not because I am boastful or vain or seek reflected glory—but because they, and all our immigrants, are the story of America. Good people, regardless of their origin, have made us a great nation. The memory of their lives can only help us be better Americans.

[signature: Henry Cisneros]

Henry Cisneros and his wife, Mary Alice, with their three children (from left): Mercedes, John Paul, and Teresa.

Secretary Cisneros meets with residents of the Arthur Cooper Dwellings public housing project in Washington, D.C., in June 1994.

Mayor Kurt Schmoke of Baltimore talks to Cisneros about urban issues affecting his city.

When Henry G. Cisneros became mayor of San Antonio, Texas, in 1981, it was a moment of triumph for the Mexican American community. He was the first Hispanic mayor to be elected in a major U.S. city.

Born in 1947 in San Antonio, Cisneros earned bachelor's and master's degrees in urban and regional planning from Texas A&M University, an M.P.A. in public administration from the John F. Kennedy School of Government at Harvard, and a doctorate in public administration from George Washington University.

He began his career in public service as an administrative assistant in the San Antonio City Manger's office. In 1971, at age 24, Cisneros became the youngest White House Fellow ever and worked as an assistant to the secretary of Health, Education, and Welfare. He was elected to the San Antonio City Council in 1975, serving until 1981, when he was elected mayor. He served four terms, doing much to rebuild the city's economic base by recruiting convention business, attracting high-tech industries, and increasing tourism and jobs. Much admired in his profession and community, he was elected president of the National League of Cities in 1985.

In 1989 Cisneros left government to become chairman of Cisneros Asset Management, which handled investments for nonprofit institutions.

In 1992, President Bill Clinton named Cisneros to his cabinet as secretary of Housing and Urban Development. Secretary Cisneros has since focused public attention on the problem of homelessness throughout the country and has worked to improve public housing, expand the supply of affordable housing, and help communities sustain economic development.

Cisneros and his wife, Mary Alice, have two daughters and a son.

Mariano Vallejo with two of his 16 children (back row), and three granddaughters. Vallejo owned an estate of nearly 250,000 acres in the Sonoma Valley when California became part of the United States in 1848. He became a state senator in the new government of California.

CHAPTER ONE

THE FIRST MEXICAN AMERICANS

The first Mexican Americans did not choose to leave their homeland and come to the United States. Instead, the United States went to them. As the Tejanos (Texans of Mexican descent) say, "We never crossed a border. The border crossed us."

The process began in 1836, when settlers in the Mexican territory of Texas—many of whom had emigrated from the United States—waged a successful rebellion against Mexico. The victorious settlers declared Texas an independent republic. Nine years later, Texas became the 28th state of the United States.

That same year, 1845, a New York City magazine editor declared that it was the "manifest destiny" of the United States "to overspread and possess the whole of the continent." The phrase "manifest destiny" became the slogan that justified the U.S. war with Mexico from 1846 to 1847. In 1848, the defeated Mexicans signed the Treaty of Guadalupe Hidalgo, giving the United States California, Nevada, Utah, and parts of New Mexico, Arizona, and Colorado. Five years later, the United States paid Mexico $10 million for the rest of what is today

Arizona and New Mexico.

More than 80,000 Mexicans already lived in this vast territory and now found themselves citizens of the United States. Many were angered by the events that linked them to a new country. The last Mexican governor of the territory of New Mexico surrendered his office to a general of the U.S. Army with the words, "Do not find it strange if there has been no manifestation of joy and enthusiasm." Mexicans regarded themselves as part of an older and more advanced culture than that of the rough Anglos who had taken control of Mexican land.

As a people, the Mexicans are a mixture of Native Americans, Spaniards, and a small number of Africans. Mexican culture is a blend of the traditions of all these peoples. During the period of Spanish rule, many Spaniards and Native Americans intermarried, producing children who were called *mestizos* (the Spanish word literally means "mixed"), who today form the majority of the Mexican and Mexican American populations. Mexicans celebrate their unique identity as *La Raza*—"The People."

Around 2000 years ago, the first major civilization arose in the Valley of Mexico, at a place called Teotihuacán, not far from today's

Mexico City. The people of Teotihuacán constructed stone pyramids and temples. Their chief god was Quetzalcoatl, a feathered serpent with the head of a jaguar. According to the legend of Quetzalcoatl, the god himself had been driven into exile by an evil magician. But one day, Quetzalcoatl would return to establish a mighty empire, a paradise on earth.

Sometime after 1100, a group known as the Aztecs left their homeland, which they called Aztlán. The exact location of Aztlán is unknown, but some have claimed it was in the southwestern part of today's United States.

The Aztecs wandered south, following the commands of their hummingbird-god, who told them to look for an eagle perched on a cactus and with a serpent in its mouth. (Today, this image appears on the flag of Mexico.) The Aztecs found the sacred eagle on an island in Lake Texcoco. There, they ended their wanderings, founding the city of Tenochtitlán around the year 1345.

The Aztecs were an industrious and skilled people, and they embarked on a series of conquests. At its height, their empire stretched across Mexico from the Gulf of Mexico to the Pacific. Influenced by the customs of the area, the Aztecs adopted the worship of

Quetzalcoatl. Indeed, their astronomers predicted the exact year of the jaguar-god's return (1519, by our calendar).

In that year, Moctezuma II, the ruler of the Aztec Empire, received reports of the arrival of strange, white-skinned men on the eastern coast of his realm. The men wore silver armor and rode on four-legged beasts that no one had ever seen before. To Moctezuma, the news indicated that Quetzalcoatl had indeed returned.

By coincidence, 1519 was the very year when the *conquistador,* or conqueror, Hernán Cortés arrived from Spain in search of glory and riches. Cortés brought a force of 600 men on horseback, armed with muskets. Within three years, the Spanish had conquered the Aztec Empire and destroyed the city of Tenochtitlán. In its place, the Spaniards would build Mexico City.

The period of Spanish rule over Mexico had begun. New Spain, as the Spanish dominions were known, would last for three centuries. During that time, the Spaniards carried out the two main goals of their conquests in the New World: spreading their religion and establishing an empire to enrich Spain.

Juan de Oñate, the husband of Cortés's granddaughter, led an expedition into the territory that is today the southwestern United States. Oñate's men crossed the Rio Grande on April 20, 1598—at a place still known as El Paso (the pass). Oñate founded the first Spanish settlement in today's New Mexico and brought the first cattle, sheep, and horses into the region.

Wherever the Spaniards went, priests of their Roman Catholic religion followed, seeking to convert the Native Americans to their faith. In Texas, the Southwest, and up the California coast, friars and brothers of the Franciscan and Dominican orders built mission churches and farms, using Native Americans as laborers. Towns sprang up around these missions,

A Californio family in the 19th century. Their clothing indicates that they were prosperous.

and the names of such modern cities as San Antonio, San Diego, Los Angeles, and San Francisco are reminders of the original Spanish settlements.

Spanish cultural traditions blended with the Native American ones. New World foods—corn, tomatoes, potatoes, and chili peppers—became popular not only in Spain but in other European countries as well. In turn, the Spanish cattle and sheep introduced milk and cheese to New World cooking. *Chile con carne* (chili peppers with beef) and the *tamale* (ground beef and chilies wrapped in cornmeal dough and steamed in corn husks) are Mexican combinations of Old World and New World foods.

The blend of traditions is probably best represented by Mexico's most popular religious devotion—that of the Virgin of Guadalupe. Near Mexico City in December 1531, the Virgin Mary appeared to Juan Diego, an Aztec convert to Christianity. The hill where this miraculous vision took place had earlier been the site of a shrine to Tonantzin, an Aztec mother-goddess. The Virgin Mary told Diego to carry flowers from the hill to the Spanish bishop in the capital. When Diego let the flowers fall from his cloak, the bishop saw the image of the Virgin imprinted on the cloth. Significantly, she had the dark skin and features of the Native Americans. Ever since, Mexicans have revered that image, known as the Virgin of Guadalupe.

Spain's control over its New World colonies began to weaken in the early 19th century. On September 16, 1810, a priest named Miguel Hidalgo rang the bell of his church in the town of Dolores. Father Hidalgo urged the assembled crowd to fight for their freedom from Spain. Hidalgo's *grito de Dolores* (cry of Dolores) was the beginning of the Mexican independence movement. Though Hidalgo's rebel army was crushed the next year, September 16 is today celebrated as Mexico's Independence Day.

The Mexicans fought on after Hidalgo's defeat, finally gaining their independence from Spain in 1821. But, as one Mexican president lamented, Mexico was "so far from God and so close to the United States." In less than 30 years after independence, Mexico had lost more than half its territory to its neighbor to the north.

The Treaty of Guadalupe Hidalgo granted the Mexicans who lived in the territory full rights as citizens of the United States. However, as Anglos (non-Mexicans) moved into the territories, this promise often proved to be an empty one.

Nine days before the treaty was signed, gold was discovered in California. Soon, a horde of gold seekers from all over the world swarmed into the region. The population exploded from about 15,000 Californios (Mexican residents of California) in 1848 to 260,000 people four years later. Californios saw the fields of their elegant haciendas, or estates, trampled by prospectors.

Among the first forty-niners to reach California were Mexicans from the state of Sonora who had experience in mining. Because they knew how to pan gold from streams and how to extract gold from ore-bearing rock, these Mexicans were often more successful than prospectors who had rushed across the continent expecting to pick nuggets off the ground. Deadly fights broke out between Mexicans and Anglos. Joaquin Murieta, a Mexican American Robin Hood, terrorized Calaveras County in the 1850s in revenge against the *gringos* (a rude term for whites) who stole his claim and killed his brother.

California was admitted to the Union on September 9, 1850. In 1851, the U.S. Congress passed a land grant act, requiring all land titles to be submitted to a board for verification. Many Californios lost their land, and even those who won their cases had to go through a long and expensive legal battle.

In the territory of New Mexico

Judge J. M. Rodríguez and his family in Texas around the beginning of the 20th century.

(which included present-day Arizona until 1862), the Mexican Americans called themselves Hispanos. Though they remained a majority of the population, their fate was similar to that of the Californios. New Mexico was officially bilingual, with both Spanish and English used in schools and the legislature, but the Anglo settlers gradually gained power. The territorial governor was appointed by the U.S. government, and only one Hispano served in this post from 1848 until New Mexico achieved statehood in 1912. By 1900, 80 percent of the former Hispano lands were in the hands of Anglo settlers, lawyers, and land companies.

Many Mexican American ranchers kept herds of sheep. When Anglo cattle ranchers arrived, the two groups clashed in range wars. The Hispano sheepherder grazed his flocks in open, unfenced pastures. The cattlemen brought in barbed-wire fencing. The fighting culminated in the Lincoln County War of 1876–78, when dozens of Spanish-speaking ranchers were killed.

In Texas, the Tejanos faced the worst treatment. The majority of people in the Republic of Texas were Anglos—settlers from the United States. Bitter memories of the war lingered, and even after statehood Tejanos were persecuted and disparagingly called "greasers" for their dark, slick hair. The Texas Rangers (the state police)—which one Tejano state legislator likened to the Ku Klux Klan, a white supremacist group—frequently killed Tejanos suspected of crimes.

Anglos were firmly in control of the Texas state government, and they cemented their power by passing laws that prevented Tejanos from voting. Only in San Antonio, in towns along the Rio Grande like Laredo and El Paso, and in some counties where Mexicans remained a majority did the Tejanos exercise any political power.

Mexicans who had been absorbed into the United States, seeing their rights and property taken away, ruefully felt themselves to be "strangers in their own land."

The Vaquero

On the large ranchos or haciendas of the Spanish domains in North America, no people were more valuable than the *vaqueros,* or cowboys who tended the cattle. The vaquero, usually a mestizo or Indian, cultivated superb skills in horsemanship and cattle handling. Their techniques were later adopted by the cowboys of the United States. Indeed, about one-fifth of the U.S. cowboys were Mexican Americans.

The high-heeled boots of the vaqueros supported their feet in the stirrups and their wide-brimmed sombreros protected their heads from the intense sun. Anglo cowboys adapted both the boots and hats. Moreover, because the cattle grazed on the vast open plains, the vaqueros needed a way to identify each rancho's animal. The branding of animals with a distinctive mark for each owner was their solution.

The vaqueros' vocabulary also became part of the language of the United States. The word *rancho* turned into *ranch,* and the word *vaquero* was garbled into the English *buckaroo.* The vaquero's prized tool was *la riata,* or rope, which became *lariat* in English. An untamed or wild horse was a *mustañero;* in English, a *mustang.* To protect themselves from the cacti and sagebrush, the vaqueros wore *chaparreras,* the English *chaps.* The vaquero's life of following the herds on the open range could be a lonely one. Often, he amused himself by playing the guitar and singing songs. The music was said to soothe the cattle, which were often disturbed by the sounds of wild animals. The vaquero sang of his life and adventures. This musical tradition was adopted by English-speaking cowboys as well.

The high point of the year for the vaquero was the summer roundup, called *el rodeo,* when the cattle were divided among their owners. For these occasions, the vaquero demonstrated his skills and daring. Taking pride in his strength as well as his style, he roped horses and steers, threw bulls by the tails, and rode both steers and untamed horses. These contests became the model for the modern rodeo.

THE MEXICAN NORTH

Between 1579 and 1581, Diego Durán, a Spanish priest, wrote a history of the Native Americans of Mexico. Durán interviewed Aztecs who told him what Aztlán, their legendary homeland, had been like.

Our forebears dwelt in that blissful, happy place called Aztlán, which means "Whiteness." In that place there is a great hill in the midst of the waters, and it is called Colhuacán because its summit is twisted; this is the Twisted Hill. On its slopes were caves or grottos where our fathers and grandfathers lived for many years. There they lived in leisure, when they were called Mexitin and Azteca. There they had at their disposal great flocks of ducks of different kinds, herons, water fowl, and cranes.... They also possessed many kinds of large beautiful fish. They had the freshness of groves of trees along the edge of the waters. They had springs surrounded by willows, evergreens, and alders, all of them tall and comely. Our ancestors went about in canoes and made floating gardens upon which they sowed maize, chilli, tomatoes, amaranth, beans and all kinds of seed which we now eat and which were brought here from there.

Aztlán lay far to the northwest of the Aztec Empire, and Spanish explorers went in search of it, roaming through what is now the southwestern part of the United States. When the Spanish founded settlements there, the vast area was named New Mexico. The Mexican Americans who built their haciendas there did not think of the region as the Southwest; it was, in fact, the far north of New Spain and later the northern part of Mexico.

Today's New Mexico attracted the first settlers from New Spain, or what is now Mexico. In 1626, Father Jeronimo de Zarate Salmerón described what life was like in the colonia, *or colony.*

Concerning the quality of the land, it is cold and healthful, with the climate of Spain. Its healthfulness is attested by the fact that the Indians reach the age of more than 100 years, for I have seen them. It is a fertile land with fine crystalline waters and much major and minor livestock is raised, and if it were not for the greediness of the governors who have taken them all out to sell, the fields would now be covered with them. A great supply of wheat and corn and all kinds of vegetables is gathered. As far as saying that it is poor, I answer that there never has been discovered in the world a land with more mines of every quality, good and bad, than in New Mexico.... We have seen silver, copper, lead, magnet stone, alum, sulphur, and turquoise mines that the Indians work with their talent, since for them, they are diamonds and precious stones. The Spaniards who are there laugh at all this; as long as they have a good supply of tobacco to smoke, they are very contented, and they do not want any more riches, for it seems as if they had made the vow of poverty, which is a great deal for being Spaniards, who for cause of greediness for silver and gold will enter Hell itself to obtain them.

Franciscan priests established missions up the coast of California. Around these missions, settlements grew. The king of Spain granted lands in the New World to colonists who established ranchos *and* estancias *(ranches and large estates) where cattle herds grazed. It was a way of life that endured into the 19th century. In 1890, Guadalupe Vallejo, a longtime resident, looked back on "the good old days" when California was part of Mexico.*

It seems to me that there never was a more peaceful or happy people on the face of the earth than the Spanish, Mexican, and Indian population of Alta California before the American conquest. We were the pioneers of the Pacific coast, building towns and Missions while General Washington was carrying on the war of the Revolution....

In the old days every one seemed to live out-doors. There was much gaiety and social life, even though people were widely scattered. We traveled as much as possible on horseback. Only old people or invalids cared to use the slow cart, or *carreta.* Young men would ride from one ranch to another for parties, and whoever found his horse tired would let him go and catch another.... Horses were given to the runaway sailors, and to trappers and hunters who came over the mountains....

Nothing was more attractive than the wedding cavalcade on its way from the bride's house to the Mission church. The horses were more richly caparisoned [decorated] than for any other ceremony, and the bride's nearest relative or family representative carried her before him, she sitting on the saddle with her white satin shoe in a loop of golden or silver braid, while he sat on the bear-skin covered *anquera* (saddle blanket) behind. The groom and his friends mingled with the bride's party, all on the best horses that could be obtained, and they rode gaily from the ranch house to the Mission, sometimes fifteen or twenty miles away. In April and May, when the land was covered with wildflowers, the light-hearted troop rode along the edge of the uplands, between hill and valley, crossing the streams, and some of the young horsemen, anxious to show their skill, would perform all the feats for which the Spanish-Californians were famous....

There was a group of warm springs a few miles distant from the old adobe house in which we lived. It made us children happy to be waked before sunrise to prepare for the "wash-day expedition" to the *Agua Caliente.* The night before the Indians had soaped the clumsy carreta's great wheels.

Santos

A new folk art arose in New Mexico at the end of the 18th century. It was called *santos* because the images of favorite saints or the Holy Family were depicted in the paintings or sculptures. Some of the artists, or *santeros,* were priests, but many were wandering craftspeople of humble origin who traveled by burro or by foot to churches and homes, crafting the image desired. In return, they might be paid in food or clothes as well as money.

The paintings, on wood panels called *retablos,* glowed with the bright colors of dyes produced from roots and plants as well as iron ore and charcoal. *Santeros* fashioned the statues, or *bultos,* by whittling the soft root of the cottonwood tree. Then the form was covered with plaster and painted in bright colors. Often, the *santeros* portrayed favorite saints from the New Mexicans' home villages in Mexico. The Virgin of Guadalupe was a favorite image.

The *santos* played an important role in the life of the New Mexicans. In each home, candles burned in front of the images as the families prayed to them for protection. The *santos'* help was requested to provide a good crop, protect a woman in childbirth, or ensure protection for the livestock. Sometimes, if the request was denied, the family retaliated against the *santos* by turning them to face the wall or even by mutilating them.

As Others Saw Them

John Bidwell, who went to California in 1841, commented on the hospitality of the Californios:

They had a custom of never charging for anything...for entertainment, food, use of horses, etc.... When you had eaten, the invariable custom was to rise, deliver to the woman or hostess the plate on which you had eaten the meat and beans...and say, "Muchas gracias, Señora" ("Many thanks, madame"); and the hostess as invariably replied, "Bueno provecho" ("May it do you much good").

Don Pío Pico, born in California, was the last Mexican governor of the province before it became part of the United States. He is seen here with his wife, Nachita Alvarado de Pico, and two nieces.

Lunch was placed in baskets, and the gentle oxen were yoked to the pole. We climbed in, under the green cloth of an old Mexican flag which was used as an awning, and the white-haired Indian *ganan*, who had driven the carreta since his boyhood, plodded beside with his long *garrocha*, or ox-goad. The great piles of soiled linen were fastened on the backs of horses, led by other servants, while the girls and women who were to do the washing trooped along by the side of the carreta. All in all, it made an imposing cavalcade, though our progress was slow, and it was generally sunrise before we had fairly reached the spring....

To me, at least, one of the dearest of my childhood memories is the family expedition from the great thick-walled adobe, under the olive and fig trees of the Mission, to the Agua Caliente in early dawn, and the late return at twilight, when the younger children were all asleep in the slow carreta, and the Indians were singing hymns as they drove the linen-laden horses down the dusky ravines.

José Ramon Pico, the nephew of California's last Spanish governor, recalled the Californios' preparations for Christmas.

There was no business in the country then but the raising of cattle. All summer long the cows fed over the oat-covered hills in the mornings and mowed away the clover into their four stomachs during the afternoons, as they stood beneath the sisal groves.

By November each one was as fat as [it] could get and yet manage to walk. The vaqueros [cowboys] and Los Indios came many days' journey across the San Joaquin Valley, driving big herds of cattle to be killed for hides and tallow. There were no roads and the only way to get the products to market was to drive the animals themselves....

In killing so many cattle there was more meat than could be eaten or sold, and choicest parts of the beef were cut into long strips, dipped in a strong boiling brine full of hot red peppers, and then hung over rawhide lines to dry in the sun, making a very appetizing and nutritious food known as *carne seca* [dried beef], which is used during the rest of the year for preparing the delicious enchiladas and chili con carne.

Soon after the first of December the last hide would be piled under the roof...the last scoopful of tallow ladled into its bag and the sacks and sacks of carne seca stored away with the chillies and frijoles [beans] in the attic.

Then commenced the preparations for the Fiesta del Cristo, la Noche Buena [Christmas Eve].

Chickens by the dozens were boiled and for days the Indians...worked pounding parched corn in stone mortars; the rich meal was rolled around the tender chicken, and the whole was then wrapped with chilies in corn husks to be steamed. These, packed away in jars, would keep until wanted by reason of the quantity of red pepper in the seasoning....

On la Noche Buena at our hacienda there were always

many families beside our own.... One Christmas Eve, I remember best, there was a full moon. Over all the ground there was a glittering frost, just enough to whiten everything, yet not enough to even nip the orange trees, which at this season of the year hang full of fruit and blossom both....

We had much music—guitars of the Mexican and Spanish type, made with twelve strings of wire, and mandolins. After supper there was dancing in the patio, coffee and cigaritos on the veranda, and singing everywhere. Someone said it was a beautiful night for a horseback ride over the valley to the Mission Santa Clara.

The horses in the corral were soon saddled. There were twenty-five or thirty of us young men and women. Our horses were the best of the big herds that were attached to every rancho.... The saddles, bridles and spurs were heavily covered with silver bullion ornaments, as in those times we put silver on our horses instead of on our dining tables; for Spaniards...live on horseback, and they eat but to live, instead of living to eat.

Riding out of the patio gate it was like a scene from the time of the Moors in Spain. As our horses snorted in the cold air they spun the rollers in their bits, making music that only the Spanish horseman knows....

Back to hacienda again faster than we came, to where the oranges leaned over the dark fountain pool. Then into the patio, where fragrant coffee from Mexico, tortillas and tamales de las gallinas [chicken] were served steaming hot from the big fireplace in the kitchen. Though we danced until morning, there was not a sleepy eye in the house. When daylight came we went to our rooms or took siestas in the hammocks under the verandas and were ready for almuerzo [lunch] when the bell rang.

And so for a week our Christmas lasted.

The Garza family owned extensive land in Texas before it became part of the United States. In 1929, some elderly Mexican Americans reminisced for an interviewer. One recalled:

The Garzas, the original owners of the land, have many descendants living in Starr County. Many of the Garzas married Texan [Anglos].... Society was different in those days to what it is now. The men were more gentlemanly, the ladies more *gentile* [refined]. The dances were held in what is now the old court house. The officers from Fort Ringgold and their wives were the honor guests. There were neither racial nor social distinctions between Americans and Mexicans, we were just one family. This was due to the fact that so many of us of that generation had a Mexican mother and an American or European father.

BUÑUELOS

In 1938, Ana Begue de Packman compiled a book of cookery customs of Spanish California. Here is her recipe for buñuelos, *which are a kind of Mexican doughnut:*

> 3 1/2 cups flour
>
> 1 tsp. salt
>
> 1 1/2 tbs. sugar
>
> 1/4 cup butter
>
> 2 eggs
>
> 1/2 cake yeast (the equivalent of 1 tsp. powdered yeast)
>
> 1/2 cup rich milk

Moisten yeast cake with warm water. Add a pinch of sugar and stir in the flour and salt and eggs to make a soft dough. Set to rise in a warm place. After it has risen, knead lightly for a few minutes until smooth. Cut the dough into small circles the size of marbles and let stand 15 minutes.

Roll each ball on a lightly floured board into a very thin pancake. Cut a hole in the center with a thimble. Fry in hot fat until puffed.

Shake together one cup granulated sugar and one teaspoon cinnamon. Dip each buñuelo in the sugar.

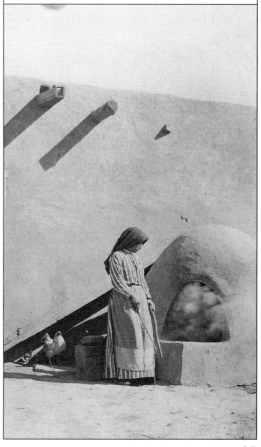

Around the turn of the 20th century, the Ontivares family gathered at their California ranch for a group portrait. They were among the Californios who managed to keep possession of their lands after California became part of the United States.

A Mexican American family in Texas around 1880. Tejanos, as the Mexican Texans were known, were often subjected to vicious prejudice by the more numerous Anglo settlers.

STRANGERS IN THEIR OWN LAND

Despite the promises made in the Treaty of Guadalupe Hidalgo, Mexican landowners in the regions taken by the United States had to fight to keep control of their property. One such Mexican complained about his situation in an 1857 magazine article.

His [the landowner's] ranch brought serious evils upon him. It was the seat of a multitude of [Anglo] squatters, who...were his bitter enemies.... They fenced in his best land; laid their claims between his house and his garden; threatened to shoot him if he should trespass on their enclosure; killed his cattle if they broke through the...fences; cut down his valuable shade and fruit trees, and sold them for firewood.

Antonio Franco Coronel was one of the Mexicans who went north to California during the gold rush of 1849. He left an account of his experiences.

The reasons for most of the antipathy against the Spanish race was that the greater portion was composed of Sonorans who were men accustomed to prospecting and who consequently achieved quicker, richer results.... Those who came later [mainly Anglo Americans], were possessed by the terrible fever to obtain gold, but they did not get it.... Well, these men aspired to become rich in a minute and they could not resign themselves to view with patience the better fortune of others....

[Coronel, his brother, and some other Mexicans staked out a claim that proved to hold considerable gold. Hearing of the success of the Mexicans, an armed group of Anglos moved in.] About ten in the morning all of these merciless people, numbering more than 100, invaded our diggings.... All of these men raised their pistols, their Bowie knives; some had rifles, others pickaxes and shovels.

Their leader...led me to understand that this [mining claim] was theirs because before we took the place, one or two months before, he with his men took possession...and that a boundary had been marked out from one side of the river to the other.... I was able to reflect for a moment that the gold was not worth risking my life in this way....

[Coronel's brother and the others wanted to fight.] My companions ran to our camp before me and armed themselves. I knew their hostile intentions. Already they had ordered that a number of horses be saddled. I arrived where they were and persuaded them to calm down. Indeed, whatever attempt they might make would be fruitless. For me the placers [mines] were finished.

Some Californios, having seen their land and homes taken by Anglos, became bandits. To Mexican Americans, such outlaws as Joaquín Murieta became heroes. Murieta described the reasons that he became a bandit.

I was once a great admirer of Americans. I thought them the noblest, most honorable and high-minded people in the world. I had met many in my own country and all forms of tyranny seemed as hateful to them as the rule of the *Gachupines* [foreigners, or Spaniards] to the Mexicans. I was sick of the constant wars and insurrections in my native land and I came here thinking to end my days in California as an American citizen.

I located first near Stockton. But I was constantly annoyed and insulted by my neighbors and was not permitted to live in peace. I went then to the placers [gold mines] and was driven from my mining claim. I went into business and was cheated by everyone in whom I trusted. At every turn I was swindled and robbed by the very men for whom I had had the greatest friendship and admiration. I saw the Americans daily in acts of the most outrageous and lawless injustice, or of cunning and mean duplicity hateful to every honorable mind.

I then said to myself, I will revenge my wrongs and take the law into my own hands. The Americans who have injured me I will kill, and those who have not, I will rob because they are Americans. My trail shall be red with blood and those who seek me shall die or I shall lose my own life in the struggle! I will not submit tamely to outrage any longer.

Josiah Royce, an Anglo prospector, wryly commented on the treatment of Mexicans during the gold rush:

We did not massacre them wholesale, as Turks might have massacred them: that treatment we reserved for the Digger Indians.... Nay, the foreign miners, being civilized men, generally received "fair trials"...whenever they were accused. It was, however, considered safe by an average lynching jury in those days to convict a "greaser" on very moderate evidence if none better could be had.... It served him right, of course. He had no business, as an alien, to come to the land that God had given us. And if he was a native Californian, or "greaser," then so much the worse for him. He was so much the more our born foe; we hated his whole degenerate, thieving, landowning, lazy and discontented race.

The Rodriguez family at their home in Bandera County, Texas. As a young man, José Policarpo Rodriguez, at left, served as a guide for the U.S. Army.

A Hispano family near Mora, New Mexico, in the late 19th century. Traditionally, Hispanos grew their crops on community farms. After the United States annexed New Mexico, however, Anglo cattle ranchers put up barbed-wire fences and seized the farmland for grazing.

This vendor sold ice cream in a southwestern town around 1890.

I have killed many; I have robbed many; and many more will suffer in the same way. I will continue to the end of my life to take vengeance on the race that has wronged me so shamefully.

In 1877, Mariano Vallejo, a citizen of California, went to Mexico City to urge the government to build a railroad from Mexico's capital to California.

I am an American because the treaty of Guadalupe [Hidalgo] placed me on the other side of the line dividing the two nations, but I was born a Mexican, my ancestors were Mexicans.... I have both Mexican and American children and I desire for my native land all the prosperity and progress enjoyed by the country of some of my children and mine by adoption. The day that Mexico has a railroad which...unites it with California, commerce and industry will progress.

Later, however, Vallejo became bitter. He had successfully defended his title to the land his family had owned for generations, only to see squatters settle on it.

The sheep introduced by the Spaniards became the mainstay of New Mexico's economy. Although the Mexican American sheepherders lost out to the Anglo cattle ranchers in the Lincoln County War, raising sheep has remained important. In the 1950s, Fabiola Cabeza de Baca described the work of her ancestors.

I can remember my paternal grandfather's sheep camps, and the men who worked for him. They were loyal people, and as close to us as our own family. They were, every one of them, grandfather's *compadres,* for he and grandmother had stood as sponsors in baptism or marriage to many of their children.

Lambing season was a trying one, since the range was extensive. This happened in the early spring, and the weather on the *llano* [plain] can be as changeable as the colors of the rainbow. If the season was rainy, it went hard with the sheep and many lambs were lost. If there had been a dry spell the year before, the ewes came out poorly and it was difficult for the mothers bringing in young lambs. Sheep raising was always a gamble until the day when feed became plentiful with the change in transportation facilities.

In order to save ewes and lambs during a cold spell, the herders built fires around the herds. The fires were kept burning day and night until better weather came to the rescue....

Shearing the sheep was done in the summer and there were professional shearers who went from camp to camp each year. This was a bright spot in the life of the herders, for then they had a touch of the outside world. Among the shearers and herders there were always musicians and poets, and I heard Papa tell of pleasant evenings spent singing and storytelling, and of *corridos* [ballads] composed to relate events which had taken place. These poets and singers were like the troubadours of old. The *corridos* dealt with the life of the people in the communities and ranches; they told of unrequited love, of death, of tragedies and events such as one reads about in the newspapers today.

The sheepherder watched his flock by day, traveling many miles while the sheep grazed on the range. As his flock pastured, he sat on a rock or on his coat; he whittled some object or composed songs or poetry until it was time to move the flock to water or better pasture.... At night he moved his flock to camp, a solitary tent where he prepared his food and where he slept. If there were several camps close to each other, the herders gathered at one tent for companionship.

In winter the sheepherder's life was dreary. Coming into his old tent at night, he had to prepare for possible storms. The wood for his fire might be wet, and with scarcely any matches, perhaps only a flint stone to light it, his hands would be numb before he had any warmth. He might not even have wood, for in many parts of the *llano* there is no wood, and cowchips had to serve as fuel.

He went to sleep early to the sound of the coyote's plaintive cry, wondering how many lambs the wolves or coyotes might carry away during the night. The early call of the turtle dove and the bleating of lambs were his daily alarm clock, and he arose to face another day of snow, rain, or wind. Yet he always took care of his sheep, and I have never known any mishap due to the carelessness of the herder....

I know an old man who worked for my maternal grandmother for many years. Often I accompanied my grandmother to the sheep camp on the Salado, and I always came back with a feeling of loneliness. Yet, at camp, the old man always seemed happy. If he was not at camp when we arrived, we found him by listening for his whistle or singing in the distance. When I think about the herders on the endless *llano,* I know that they are the unsung heroes of an industry which was our livelihood for generations.

In 1991, at the age of 84, Sophie Rodriguez Waldrip recalled the history of her family in New Mexico. Her stories, including tales she had heard as a child, extended back in time over nearly 150 years.

My great-grandfather, José Maria Ramirez, was one of the very early settlers in the San Miguel, New Mexico, area which, at that time, was part of the Territory of New Mexico. He married my great-grandmother, Doña Iñez, who was part Indian...of the Apache tribe. They had three children, two girls and a male. I remember that my great-grandmother was small, and dark-skinned, with white hair. In 1920, she died at the age of 106. She never complained about an ache or a pain and never wore eye glasses. The last time I saw her was a day when the sun was shining so pretty and I stopped by to talk to her. She was in the front part of the

Mexican Americans sometimes organized widespread resistance against the injustices of Anglo laws. In the late 1880s, a group of native-born New Mexicans formed a secret society called the Gorras Blancas, *or White Caps. Using guerrilla tactics, they cut down barbed-wire fences, destroyed railroad tracks and telegraph lines, and burned Anglo ranches. On the night of March 11, 1890, the White Caps posted copies of their program in public places in the town of Las Vegas, New Mexico:*

Our purpose is to protect the rights and interests of the people in general and especially those of the helpless classes....

We are not down on lawyers as a class, but the usual knavery and unfair treatment of the people must be stopped....

Our judiciary hereafter must understand that we will sustain it only when "Justice" is its watchword....

We favor irrigation enterprises, but will fight any scheme that tends to monopolize the supply of water sources to the detriment of residents living on lands watered by the same streams....

We must have a free ballot and fair court and the will of the Majority shall be respected.

We have no grudge against any person in particular, but we are the enemies of bulldozers and tyrants.

If the old system should continue, death would be a relief to our suffering. And for our rights our lives are the least we can pledge....

The White Caps,
1,500 Strong and Gaining Daily

Guadalupe Triviz Chavez with her children in the 1920s. Born in Mexico in 1883, she married Epefaniro Chavez, whose family owned a farm across the border, and moved to New Mexico, where her family lives today.

Juana Elias, a Mexican American girl in Arizona, dressed for her First Communion in 1901.

house sitting by the window in the sunlight mending my great-grand-uncle's socks.

My great-grandfather Ramirez built his house in San Miguel about 1854. All the neighbors came to help, like they did in those days.... The house has been a haven for some member of the family all these years and is now my home....

Social life at San Miguel in the 1860s up to the early 1900s as I remember hearing about it, centered around my great-grandmother Iñez's big living room.... Whenever the people living in the area had any kind of a meeting, a dance, baptism, or other special occasion, they used that room. They were mostly Catholic families but their small adobe church had no hall. I remember that room had a mural painted on the back wall. It depicted the life of Christ and the saints. I wonder now why we didn't leave that mural. It was fading away, but we could have had it touched up....

The fiesta and all the community gatherings were always held in my great-grandmother's living room until 1912, when the State of New Mexico became part of the Union. That year the state allowed some funding to build the first public school in San Miguel. After that time, the community used their new three-room white brick school building for their meetings.

Sophie Rodriguez Waldrip also remembered her mother's mother, who was a true pioneer wife.

Her name was Margarita Enriguez. Friends and relatives called her Grandma Lita. She married my grandfather, Don Eugenio Moreno, when she was fifteen and he was sixty-five. It was a pre-arranged marriage.... Grandma Margarita and Grandpa Eugenio had ten children.

[After her husband became ill] she managed the farm and the workmen.... In those days they raised mostly wheat and some corn. In later years the farmers in the valley raised grapes for wine, and grew peaches, apples, cherries, pears, plums and strawberries. They had some irrigation but when the rains came, and they did come, San Miguel would be flooded and the farmers would lose their crops and have nothing to harvest....

Grandmother was always very busy. In fact, she was a midwife, and when people would call on her to deliver a baby, she would have to make arrangements to have someone take care of the farm while she was gone. She used to deliver babies from near El Paso to Old Mesilla. She was the only licensed midwife in those days, since there were no rural doctors.... All through the valley Grandma was known as La Señora Doctora. She used to say, very proudly, that in all the valley, she never lost a baby or a mother. They would come in a wagon to pick her up....

Grandmother Margarita also took care of sick people all over the countryside. She used various kinds of herbs as medicine for pneumonia, colds, and different ailments that people developed. She had chamomile, which was good for many things, *Cascara Sagrada* [sacred hide], a laxative and good for digestion, *Gordo Lobo* [fat wolf] for colds, coughs, and asthma.

In the 1980s, Elina Laos Sayre remembered her life as a girl in Tucson, Arizona, in the early years of the century.

When I was a young girl we would go to the Placita where the first St. Augustine Cathedral was. There was a gazebo in the center of the plaza and a string orchestra played there. The gazebo was so beautiful. It was very ornate. It had a roof over it and it had pillars and all around there was interlaced woodwork. And it had steps going up to it. There was grass all around it and great big boulders that we used to sit on...and listen to the orchestra.... When the boys came we would get up and parade on the inside circle around the gazebo. And the boys would make another circle and parade on the outside in the opposite direction. And the first thing you knew, the boys would be intermingling with the girls. We would be all dressed up with long beautiful dresses and we wore hats with feathers or flowers....

Sometimes on Sunday we would go for a *barbacoa* to the Amado Ranch out on the Nogales Highway. We would all go in a wagon pulled by farm horses.... Before we got to the ranch we could smell the wonderful aroma of the barbecue cooking. They used to dig a big pit to cook the meat in. There was a long table with a white cloth. There was always so much food. Big pots of beans and green chile salsa and corn, and stacks and stacks of tortillas. And then we would dance...waltzes and mazurkas.

Those were beautiful, romantic days.... They were full of happiness, love, music and *abrazos* [embraces]. We all loved one another and helped one another the best we could. Sometimes...when I get tired of watching TV or reading, I just go to bed and reminisce. I have had a wonderful life. I am very lucky.

Federico Ronstadt's Club Filarmónico around 1895.

The Ronstadt Family

In 1882, when he was 14, Federico Ronstadt left Mexico and settled in Tucson, Arizona. He worked as an apprentice carriage maker. By 1888, he was successful enough to start his own carriage business. He was a respected business leader in Tucson, serving as head of the city's Chamber of Commerce.

Ronstadt also organized a musical group with other young Mexican Americans. He recalled: "We started with eight or ten members—some of them knew a little about music but the others didn't know a note. It was up to me to teach them what I could.... Dick [his younger brother] played the flute and I learned the fingering of the clarinet. The rest had to learn violin, viola, cello, bass, trombone and cornet." The group named themselves the Club Filarmónico and started playing at a local restaurant. In the summer of 1896, they toured Southern California.

Ronstadt's daughter Luisa, born in 1892, carried the family tradition a step further. Under the name Luisa Espinel, she became an internationally famous singer, performing Hispanic folk music in Los Angeles, New York, and Europe. Later, she appeared in movies as well. After retirement, she settled in Los Angeles, where she recalled her childhood: "There were...summer evenings [when my father] would accompany his songs on his guitar and later tell us marvelous stories of when he was a little boy.... The most vivid memories of my childhood are interwoven with music and mostly the music of my father, who loved it."

Today, Linda Ronstadt, Federico's granddaughter, has become the most famous member of this musically talented family. She burst onto the rock music charts with her hit record "Different Drum" in 1967. At the time, she declared her intention to be "the world's greatest Mexican singer." In the 1980s, she toured the United States with a traditional Mexican mariachi band led by Rubén Fuentes. Ronstadt titled the album she made with Fuentes's group *Canciones de mi Padre* (Songs of My Father). Her father, Gilbert, still ran the family hardware store when Linda was growing up in Tucson. She would accompany him on trips to Mexico, and whenever they crossed the border, he would throw back his head and sing. Though Gilbert Ronstadt was never a professional musician, he passed on the family gift to his daughter.

A family in Mexico in the early part of the 20th century.

CHAPTER TWO

GOING NORTH

hen Mexico became an independent nation in 1821, it had about the same amount of territory as the United States did. In addition, Mexico was blessed with natural resources— gold, silver, oil, timber, a long coastline on two oceans, grasslands that provided grazing for cattle and sheep, and fertile soil.

Yet Mexico has not attained the same degree of prosperity as its northern neighbor. In the 20th century, several million Mexicans have come to the United States in search of work and better lives. The loss of more than half its territory is one reason for Mexico's comparative poverty and its failure to provide sufficient opportunities for its people. However, the country's troubled history contains other reasons.

The newly independent nation of Mexico faced serious problems. The mining industry, which was the chief source of its wealth, had been destroyed in the struggle for independence and did not recover for years. Most of the land remained in the hands of *ricos* (the rich) and the Catholic church, which was exempt from taxation. The new government had to bor-

row money from foreign bankers to meet its needs, but this dependence on outside sources of funds gradually weakened the country.

Furthermore, though Mexico became a republic, the nation's military leaders frequently overturned elected governments to take control. The most notorious of these military strongmen was Antonio López de Santa Anna, who dominated the country's politics between 1833 and 1855 in spite of an unbroken record of failure and corruption.

It was Santa Anna who personally led an army into Texas to put down the rebellion there in 1836. On March 6 his forces captured the fort at the little mission church known as the Alamo. Santa Anna gave the order to take no prisoners—an action that resulted in the Texans' battle cry: "Remember the Alamo." Only a few months later, Santa Anna blundered into a trap. His army was destroyed at the Battle of San Jacinto and he himself was taken prisoner.

Unfortunately for Mexico, the Texans released him after their struggle for independence was won. In 1846, Santa Anna once more was called on to lead his country's forces in the U.S.–Mexican War. His military skill had not improved, and he was forced to flee Mexico City as U.S. troops

marched into the capital.

Amazingly, Santa Anna returned from exile in 1853 to become dictator yet again. By this time, however, Mexicans had become convinced that the country needed serious reform. After Santa Anna squandered the country's treasury, he could no longer pay his officers, and a rebellion of liberal leaders overthrew him in 1855.

Among those who helped depose Santa Anna was Benito Juárez, a full-blooded Native American who became one of Mexico's most beloved leaders. As minister of justice in the new government, Juárez issued a law (the *Ley Juárez*) that abolished the special privileges granted to army officers and the clergy. After Juárez became president, he went even further, confiscating the church lands and putting them up for sale.

In 1861, Juárez suspended all payments on Mexico's large foreign debt so that the government could devote its funds to improving the country. This move brought an invasion of French, British, and Spanish troops to enforce their bankers' claims on Mexico's government. Juárez negotiated an agreement with the British and Spanish. But the French emperor, encouraged by Mexican opponents of Juárez's reforms, ordered his forces to march on toward Mexico City.

The Mexican army fought bravely against the foreign invaders. On May 5, 1862, under General Ignacio Zaragoza, the Mexicans defeated a French army at the city of Puebla. The anniversary of this victory, known as Cinco de Mayo, has been celebrated ever since by Mexicans and Mexican Americans. Along with September 21, Independence Day, it is one of the two greatest political holidays in the Mexican calendar.

Cinco de Mayo, however, was only a temporary victory. The French captured Mexico City in June 1863, and Juárez fled to the northern part of Mexico.

The French sent an Austrian prince, Maximilian, to be the emperor of Mexico. Meanwhile, Juárez traveled from town to town in his black coach, organizing resistance. Maximilian proved to be a well-meaning but ineffective ruler, and after the U.S. Civil War ended, the United States forced France to remove its troops. In May 1867, Juárez's forces captured Maximilian and executed him. Mexico became once more a republic.

Juárez served as Mexico's president until his death in 1872, and he struggled to solve the country's problems. He began to build railroads to connect the country's isolated northern areas with the more densely populated central region. He established a school system to educate the illiterate Native Americans and mestizos who scratched out a bare existence on the land.

The Native Americans had traditionally farmed large communal plots of land, called *ejidos*. The government broke these up to encourage private ownership of land. But the smaller plots were not economical, and the Native Americans

In Puerto Colorado, Mexico, in 1985, Doña Manuela Estrada Ramirez holds her youngest grandson in front of the homemade shrine that reflects her deep religious devotion. She has 15 children, many of whom work as migrant farm workers in the United States.

and mestizos who had received land grants sold them or were cheated out of them. To develop industries, mines, and railroads, Juárez was forced to borrow money from foreign banks and to allow foreign corporations to establish businesses in Mexico.

All these policies continued during the long reign of Porfirio Díaz, the dictator who ruled Mexico from 1884 to 1911. A mestizo who grew up in poverty, Díaz established alliances with church leaders and the landowning *ricos*. He turned to the United States as the chief source of support for Mexico's development. U.S. investment in Mexico exceeded $1 billion by 1910.

In some ways, Díaz's policies helped the country. During his regime, about 15,000 miles of railways were built, textile industries were developed, and mines again began to produce valuable silver and copper. Oil was discovered in Mexico, and the country became one of the world's largest producers of petroleum. However, the foreign owners of these enterprises exploited their Mexican workers, while the government cooperated by crushing labor movements.

The railways made it easier for people to move from the more heavily settled central plateau (where the capital is) to the less developed northern regions. Unfortunately, this freedom of movement only encouraged immigration across the border into the United States.

Little of the wealth produced by mines, industry, and agriculture trickled down to the average Mexican. By 1910, more than 95 percent of the rural population consisted of landless *peons* (day laborers). It was easy for many of them to cross the border into the United States in search of jobs and higher pay. The Rio Grande—which in many places is a shallow, slow-moving stream that even children can wade across—was no obstacle to those seeking a better life.

North of the border, Mexicans quickly found work, for labor was in demand on the U.S. southwestern frontier. In the 1880s, experienced Mexican railroad workers were hired to lay tracks between Texas and California. Completion of this railroad was a great boon to the economy of the Southwest. The growing prosperity of the area created more jobs. Farms required laborers to plant and harvest the crops. Ranches needed people to tend the herds and drive them to the railroad. Mining companies depended on the hard labor of men who were willing to do the difficult and dangerous work. Mexicans were willing to perform all these tasks, and no one asked which side of the border they had come from.

Meanwhile, in Mexico, conditions grew worse. In 1910, when President Díaz was 80, the long-simmering resentment against his rule broke into rebellion. The revolutionary movement was led by Francisco Madero, a *rico* who believed that Díaz had allowed foreigners to acquire too much power in Mexico.

Madero won support from two charismatic leaders of humble backgrounds: Pancho Villa and Emiliano Zapata. Villa, formerly a bandit chieftain in the north, and Zapata, a skilled vaquero in the south, both formed guerrilla forces that fought Díaz's armies. Zapata was eloquent in both Spanish and Nahuatl (the language of the Aztecs), telling his followers, "Seek justice not with your hands, but with a rifle in your hands."

Villa's forces captured the city of Ciudad Juárez on May 20, 1911, as the residents of neighboring El Paso, Texas, watched from their rooftops. Ten days later, Zapata's army swept into the city of Cuernavaca. News of the victories spread throughout Mexico, and other bands of rebels arose. Sensing that his time had come, Díaz went into exile, and Madero

Three rebel soldiers who fought in the long and bloody Mexican Revolution that began in 1910.

became the president. But the revolution was far from over.

Zapata demanded that Madero divide up the *ricos'* large landholdings and distribute them among the peasants. Madero, himself a hacienda owner, wanted to proceed gradually toward land reform. Disgusted, Zapata proclaimed his own program, the Plan of Ayala, which demanded one-third of all the land for the peasants.

Madero was caught between the conservatives who merely wanted a change in leadership and the radicals who fought for social reforms. In 1913 General Victoriano Huerta

took power in a military coup and executed Madero.

Civil war broke out all over Mexico. Governors of some states refused to recognize Huerta's regime and formed their own armies. Huerta was driven out, but the country remained in turmoil; the soldiers of both sides roamed the land, leaving devastation in their path.

The people who suffered most were the peasants, who saw their fields trampled, their homes burned, and their few possessions stolen by soldiers. Young men were often rounded up to serve in the armies that passed through their villages.

The revolutionary violence dragged on until 1920. During that time, more than a million Mexicans died, and the fighting created enormous damage to property as well. The resulting misery caused a great number of refugees to flee across the border into the United States. By some estimates, almost 10 percent of the population of Mexico emigrated—to the United States and other countries—between 1910 and 1930.

The agriculture and industry in the southwestern states of the United States could not absorb all of the new arrivals, and many Mexican immigrants started to move farther north in search of jobs. They brought along their customs, food, religion, and their memories of the turbulent history of Mexico. A new chapter in the story of the Mexican Americans was beginning.

This map shows the territory that Mexico lost to the United States between 1836 and 1853. Today, Mexican immigrants, legal and illegal, enter the United States through the twin cities along the border.

At the beginning of the 20th century, the great majority of Mexicans were living in poverty. Though the Mexican Revolution began as a struggle among the upper classes for control of the government, poor people like these rallied to the support of popular leaders, such as Emiliano Zapata and Pancho Villa, who wanted to bring about real economic change.

LIFE IN MEXICO

In his autobiography, Ernesto Galarza described his childhood in the town of Jalcocotán, Mexico, around 1900.

Jalcocotán was for the most part an easy place in which to live.... With my two cousins and other boys of my own age I always had something to watch or to do.

The near side of the pond was shallow and fringed with reeds and tall clumps of grass that blossomed with plumes of cream-colored fluff. Around them the pond was always muddy and cool. In your bare feet you sank up to the ankles and by wriggling your toes you could raise oozy, iridescent bubbles....

Water snakes were everywhere, which we imagined were poisonous *víboras* or copperheads, like those the *jalcocotecanos* [the people of Jalcocotán] found in the forest. We spiced our play with a legend about an alligator that had crawled all the way up from Miramar and lay in wait for us somewhere in a deep pool of the arroyo—a monster no less real because he lived only in our imagination.

When the older boys of the village came to the pond on Sunday afternoons we watched them swim and dive. From a high branch of the big *nogal* [walnut tree] they dropped a swing made of bush vines we called *liana*, braided like the women of the pueblo did their hair. The boy who was to dive next waited up in the *nogal*. Another handed him the end of the *liana*. The diver kicked off and let go as high as he could swing, his naked brown body twisting through the air like a split string bean. On our side of the swimming hole the smaller boys stripped and paddled while the divers yelled instructions on strokes and kicks.

Once in a great while the older boys would also allow us to join them in the bullfights they organized in one corner of the pasture....

From behind a tree a trumpeter stepped to the edge of the ring. Blowing on a make-believe bugle he sounded a call and the bull rushed in—a boy with a plain sarape over his shoulders, holding with both hands in front of his chest the bleached skull of a steer complete with horns. Between the horns a large, thick cactus leaf from which the thorns had been removed, was tied. It was at the cactus pad that the matadores and picadores aimed their wooden swords and bamboo spears.

As Galarza explained, the feminine and masculine roles were strongly defined.

The girls of the village, needless to say, did not take part in the swimming parties or in the action of the bull fights.... Little by little [we boys] noticed that only girls had their earlobes pierced, wearing bits of string until their parents could afford genuine rings. They had to sit for hours to have their hair braided. At five years of age girls began to learn to carry water up from the arroyo in *ollas*, holding them on top of their heads with both hands, something that no man in Jalco would think of doing. They played silly games like La Ronda, hopping around and around, we thought, like *zopilotes* [buzzards]. Boys did girls' chores only if, and everybody knew that it was only if, there were no girls in the family—like shaking and sunning the bed mats or sprinkling the streets in front of your cottage.

Sotero H. Soria, who immigrated to the town of Wellington, Kansas, described what it was like to grow up in a Mexican family.

The parents were very dedicated to their families. At a certain age, they began to inculcate into their children respect, obedience, and humility.

The father was the supreme authority in the family. When the children began to grow older, the father took charge of the sons and taught them their duties. The mother taught the

Shoppers stroll through Market Square in Mexico City around 1905. Farmers from outlying regions customarily took their crops to public markets in towns and cities throughout the country.

Father Miguel Hidalgo

On the morning of September 16, 1810, Father Miguel Hidalgo rang the church bell in the town of Dolores, about 120 miles north of Mexico City. The bell brought people running from throughout the town. Father Hidalgo spoke eloquently to his parishioners: "My children, will you be free? Will you make the effort to recover from the hated Spaniards the lands stolen from your forefathers three hundred years ago?" He urged them to take up arms and follow him, ending his speech with "Long live Mexico!" Father Hidalgo's speech is known as the *grito de Dolores* (cry of Dolores) and is celebrated as the start of the movement for Mexican independence.

Father Hidalgo's flock, armed with pitchforks and shovels, began to march toward a neighboring village. On the way they stopped at a church and took a banner of the Virgin of Guadalupe as their standard. As the ragtag band moved through the countryside, new followers swelled their ranks.

By the time Hidalgo drew within sight of Mexico City, he had more than 80,000 volunteers behind him. On a hill outside the city, the rebels won a fierce battle with a force of Spanish soldiers. But this fight was a costly one. Because his ammunition was low and his followers exhausted, Hidalgo decided to withdraw to Guadalajara to give his army time to regroup.

It was a crucial mistake. A larger army of Spaniards pursued and crushed the rebel army. Father Hidalgo fled with some of his followers but was captured. On July 30, 1811, he was led before a firing squad in Mexico City. He blessed the soldiers who were to execute him, promising them that Mexico would one day be free.

Although the Spanish did not relinquish control of the country for another ten years, Father Hidalgo is honored today as the father of Mexican independence. The anniversary of the *grito de Dolores* is still a day of celebration in Mexican communities in both the United States and Mexico.

Children in a Mexican village had many ways to amuse themselves, as Sabino Martinez Arrayo recalled:

A large group of boys used to get together in the fields. There was corn, mesquite, and pumpkin seeds that we roasted and ate. We also always roasted chickpeas.

We used to go there near the hill where there were many edible fruits. On the trees were fruit such as guavas and sapotes. There were several types of prickly pear, the white ones and the gentle ones. There were strawberries and a very tasty herb.

Also, we ate a lot of cane that is used to make sugar. We used to go to the cane fields to help the planters. There was a lot to eat if one offered his services for nothing....

On the hill where the animals wandered loose, we rode them like cowboys do. Because we were very mischievous with them, now and then they threw us off. We spent the time very happily and had a lot of fun when we weren't working.

A candle maker displays his wares. Skilled craftspeople sometimes raised enough money to open a shop.

daughters their domestic tasks. The parents were always concerned that their children be honorable.

After dinner we used to ask permission from our parents to go play with the neighbors. Our games were simple, happy ones.

In one of the games, The Hens and The Coyote, a group of boys and girls held each other by the hand. One, the coyote, tried to grab one of the hens, but the rest of us kicked him so that he wouldn't get her. Finally, the moment arrived in which he grabbed one and carried her off.

Naturally, when they reached a certain age, the sexes were separated. The young women went with young women, and the young men with young men. Fathers and mothers were always very suspicious.

When the children decided to marry, the parents gave them good advice. For example, the mother always instructed her daughter to be a good wife and to obey her husband. Their mothers tried to make sure that their daughters were dressed decently and modestly at church and fiestas.... Fathers advised their sons to show respect to the young women.

Then, a young man was not permitted to smoke in the presence of older persons, not even his parents. Also, if, for example, someone older asked for a glass of water, a youth would bring it to him, remove his hat, and wait until the glass was returned to him. Youths avoided offensive language in front of their elders. These were signs of respect that were transmitted from generation to generation.

My wife and I feel very fortunate that we were brought up as we were.

Margarita Avila grew up in Mexico in the 1920s. Years later, when she was living in Los Angeles, she recalled her middle-class upbringing.

My mother had a very pretty voice, and she used to sing operas or operettas, and my father would sing back to her; they used to talk to each other that way. Sometimes they would be singing to each other and my mother would sit on my father's lap. They loved each other so much and communicated with each other so well that it impressed me when I was a child.

My father was a hard worker; he had no vices. He had a taxi when we lived in Chihuahua, and we lived very well. I remember because my sisters and I were the best-dressed children in school. We were five children, and I remember my mother had somebody to help her.... I belonged to a social club. In those days, social clubs were all about parties and more parties. With all I had going on in my life, I didn't need too much more. Because my father had been liberal in terms of our education, we learned sports such as bicycle-riding, which were usually for boys. But my father was progressive in that way....

For a young person I think my era in Mexico was the best. There wasn't so much familiarity then; boyfriends didn't come to the house so much. It was harder to see each other. And there was more opportunity to flirt with other people. I think it was all so much nicer then, maybe because I was a romantic. [She turned down a marriage proposal.] He wanted to get married, but I wasn't ready. I wanted to know what was going on in the world, and I thought I'd get married when I was thirty years old. That was unusual for those times; my little sister thought she was a spinster when she was twenty-one.

Cesar Rosas, who is the lead guitarist with the popular music group Los Lobos, recalled his boyhood.

When I was a little kid in Mexico I remember the Saturday night dances in the ranch. People would come from different areas. I'll never forget that, my mom and dad dancing. There was no electricity, right, so they'd bring generators and hang strings of light bulbs, socket and extension cords, and all these bright light bulbs. That was the dance area and you'd have the Norteño [northern] groups. The Norteño guys, the *bajo sexto* [bass guitar], and the accordion doing all these great songs of the time. I've been collecting that kind of music maybe ten or fifteen years. I would buy records and play them, "Man, I remember this song." When I was a little kid this song was a hit. These guys were playing underneath those lights over there. You know it was incredible....

My dad was a diesel mechanic. My dad was born in a little town called La Purisima, Baja California, way down near La Paz.... During that time, there was an American, a rancher who owned a lot of land in Hermosillo. So he hired my dad to take care of his ranch. He took my mom there. When they got there, we started being born. There aren't many of us. The only ones living now are my older brother Pete, then it's me, then Rudy is

In this small Mexican village, the feast day of the patron saint of the local church gave people a chance to enjoy themselves and take a break from their everyday chores.

Corn and potatoes, two of the world's most important crops, were developed by Native American farmers. This woman, in the Yucatán Peninsula in the 1980s, carries corn husks left over from the harvest.

younger, and Patti is the youngest. We lost a little brother between me and Pete. Back in the fifties, in Mexico, it could get really tough. A lot of poverty. Not only that, all the odds were against us because we were isolated, living in a little ranch with no electricity, no running water, no indoor plumbing. When you're a little kid you're just happy to be anywhere. I saw a lot of hardships, but I never connected it to anything. You sense that there's something wrong, but you just go on being a little kid, being happy with whatever you have. Oh, they struggled so much.

Dolores Rodriguez King described the Christmas celebrations in her parents' home village in Mexico.

Where my parents were born, in preparation for Christmas festivities, it was the custom to celebrate the birth of Christ with songs of praise.

My mother told us that, for the Catholic Church in Mexico, those were the days of greatest importance. From among the people of the hacienda on which she lived were chosen the Magi, the angels, the Virgin, and the Christ Child.

During the nights of the Posadas [nine-day celebration of Christmas], everyone gathered at the edge of the hacienda and singing, went to the first house. There they asked for lodging but those within replied that they had no room for the Christ Child.

From there they proceeded, still singing their songs of praise. Everywhere they stopped, they received the same answer to their request as they had at the first house. Finally they arrived at the place that had been prepared to receive the Christ Child.

In that house, everything was now ready for the celebration. The parties began before Christmas and continued until January 6th. Each night there were dances, food and songs.

When he came to Garden City [Michigan], my father's uncle organized Posadas there. For several years they continued to celebrate them as they had in Mexico.

Many devout Mexicans saw the hand of God in all the natural phenomena around them. Maria Perez Rivas, a 20th-century immigrant, recalled:

When there was a storm, my grandmother used to say, "Come, come bring a knife to cut the cloud."

We stopped what we were doing and blessed the storm. "In the name of the Father, of the Son, and of the Holy Ghost, the Blessed Trinity, and Jesus, Mary, and Joseph, free us from this storm."

And the storm abated. But the one who had to do it must not be a sinner. So, when we were very young, we were the ones chosen.

Once, there was a furious storm and my uncle, who was still very young, took off his sandal, blessed the cloud with it, threw his sandal into the air, and the storm calmed down.

Some Mexicans performed pilgrimages to religious sites to obtain favors, as Juanita Silva of Manhattan, Kansas, recalled.

Another pilgrimage was to visit the Basilica of Our Lady of San Juan de los Lagos, the famous and miraculous Virgin of the State of Jalisco.

People went on the pilgrimage to fulfill the vows made to the Virgin when they had first petitioned her. If the Virgin complied with their petitions, they returned to the church to pay their obligations. Some entered on their knees to visit her. They brought her candles or flowers.

At times the trip on foot was so arduous that often the people tired and refused to continue. And that decision boded ill for those who made it because they were turned into stones. It was said that the rocks found on the road were large ones.

Sabino Martinez Arrayo remembered how his grandmother in Mexico used herbal remedies to treat injuries.

My grandmother, who was by now an old woman, treated others. I used to bring her plants that grew on the hills: hyssop, savin, terebinth, rhubarb, aloes. In the town where I was born and raised, there was no doctor. She ministered to people who were bruised or hurt with herbal medicines. She served as a doctor, although she had no formal training, to women about to give birth. She did this very well for there were some women with five or six children who never complained of having suffered any after effects.

She knew many medicines. She said that God taught her to recognize herbs that could help others.

Sabino Martinez Arrayo, a 20th-century immigrant, described the customary preparations for marriage in her home village. First, the groom asked the local priest to serve as a go-between. Then:

The priest of the church asks for the bride's hand. If she accepts, the bride's father requests six months time to prepare her.

The bride does not spend those six months in her father's home. Rather she stays at the home of some friends of hers. And the groom has to go by daily as if she were already his wife and he were supporting her.

On the wedding day, the couple is served a breakfast of chocolate, Mexican bread and some flavorful drinks.

After they are married and a child is born, they seek godparents who will take it to be baptized. The godparents must buy the clothing for the child and then throw a party with a lot of food. At times, if the godfather is well off, he holds a dance also.

When they leave the church, they throw money, which they call *el bolo,* to the youngsters who are waiting. They throw it into the air, and the children run to the street to collect it....

It was also the custom that if the parents died, the godparents had the right to adopt their godchildren; thus, they took the place of the father and mother.

A Mexican wedding party in Ciudad Juárez around 1910.

Virtually all the possessions of this family, photographed in the Yucatán Peninsula around 1900, were handmade. The development of Mexican mines and industry, financed by foreign companies and banks, did little to improve their lives.

THE DECISION TO LEAVE

For many Mexicans, fleeing to the United States was their only chance to escape a life of poverty and hardship. Long after he came to the United States, Gregorio Mujica remembered an incident during his boyhood in Mexico, when the dictator Porfirio Díaz ruled the country.

Porfirio Díaz' government treated us like animals on the haciendas. There were schools for those who were somewhat protected by the rich, but the poor were ignored.

When I was a little boy, I started out earning twelve centavos a day. Some fifty or sixty of us, under a foreman, worked at bundling wheat. We would run with the bundles to throw them in some wagons. When one was fairly well filled, a large man would stoop down and toss the bundle to another man who was on top of the wagon. If the boys weren't fast enough, the foreman beat on their feet. It was a bitter slavery.

I was arrested when I was a boy because I didn't want to work at dreadful jobs that I couldn't do. I was coming from the fields when the boss' shadow, Rafael Andrade, grabbed me and arrested me. For three nights I stayed in jail. The man who later would become my stepfather, came to stay with me after he got off work.

One day I went out and sat in the doorway at the side of a guard.

I heard him exclaim: "There comes the boss, one of the great ones, and now you'll see. He's going to have it out with me."

"Hey, lad, what are you doing there?" the boss asked me.

"Well, sir," the guard answered, "he's a prisoner. We only let him out as far as the door."

"That boy, a prisoner?"

"Yes."

"Who told you to arrest him?"

"Don Rafael brought him."

"Well, let him go home."

They released me.

Sotero H. Soria also recalled the harsh treatment that the overseers on the haciendas gave to the child laborers.

The period of the great landowners was for the poor Mexican people a dreadful one. They worked from dawn to dusk for wages that were not sufficient to feed their families.

I remember that when I was some six or seven years old,

they forced us to reap the wheat and put it in the carts drawn by oxen.

The ranch foreman arranged us in groups, and with the whip that he always carried, flogged each of us. He told us that it was to keep us warm. Naturally, it warmed us up. Screaming with pain from the lashes, we jumped and cavorted like goats. How could we help but run to work with such fear?

I would like to forget those times.

The outbreak of the Mexican Revolution in 1910 began a long period of fighting in which many different sides fought for control of the country. Those who suffered most were ordinary people caught up in the violence of the time. Miguel Garibay of Manhattan, Kansas, described how his father was almost shot by a band of revolutionaries.

When the Villistas, the Carranzistas, or the Maderistas came to a town, some acted like bastards. Since many of them were from the lower class, there was no order. If they wanted a woman, they carried her off by force. When they entered a town they demanded money, women, horses, and pistols.

They allowed the poor people to go into the shops and take whatever they wanted—clothing, corn, or anything else. Many indeed took what they wanted. Later the government men came and said that they had to hand over what they had robbed. By then the revolutionaries had left....

At the time of the Revolution my father owned a shop and a bar. He would go to Irapuato to buy his wares. Back then, there were no trucks, or if there were, they were few in number. Therefore, they used to go in a small stagecoach which ran from Cueramaro to Irapuato.

On the way there, they were held up by two men; they were probably either revolutionaries or scoundrels....

They asked my father his name. When he replied that it was Antonio Garibay, they reacted violently.

"He is one of those who shot at us from the church tower," they said. "Let's hang him."...

Right away they carried my father to a tree. They took off his shoes and all his clothing except his shorts.

My father asked them who their leader was. It was a man who was a tailor. Since my father was also a tailor, he knew him. He insisted that they take him to see their leader. Finally they did.

"How are you doing, Antonio?" the leader greeted him.

"Well, they are going to kill me," he answered.

"Why?"

When the leader learned the details of the situation, he told his men to free my father and give him back his clothing. He arrived back home with everything but his shoes. One of the bandits had gone off with them. What a close call he had!

Pancho Villa

One of the most colorful leaders of the Mexican Revolution was Doroteo Arango, better known as Pancho Villa. Born in the northern Mexican state of Durango in 1878, Doroteo's parents were poor sharecroppers on a large hacienda. From childhood, the boy was rebellious. After a wealthy rancher raped Doroteo's sister, the young man killed him. Doroteo was imprisoned for the murder but soon escaped. At this time, he took the name Pancho Villa, after an outlaw ancestor.

The outbreak of the Mexican Revolution gave Villa the chance to avenge all his grievances against the rich. He formed a rebel army and in 1911, his troops captured Ciudad Juárez, making him a national hero.

The rebel movement was disorganized, and armies like Villa's followed no orders except their own commander's. Villa attacked the government troops, or *federales,* in hit-and-run raids. One Villa follower remembered, "We learned to ride like hell, to eat when we had food, and to sing when there was none."

In 1914, Villa joined forces with the most important rebel leader from the south, Emiliano Zapata, and captured Mexico City. However, their victory was short-lived, for a better-equipped army under General Alvaro Obregón drove them out of the capital. Villa retreated to his northern stronghold.

Villa became angered when the U.S. government supported one of his rivals. In retaliation, he led his forces across the border, shooting up the town of Columbus, New Mexico, in 1916. This was the only time in the 20th century that the United States was ever invaded. U.S. President Woodrow Wilson sent a military force under General John Pershing into Mexico to capture Villa. But the rebel knew the territory too well, and Pershing never caught him.

The fighting of the revolution died down after the Constitution of 1917 was written. Two years later, the new Mexican government offered Villa amnesty and granted him an estate. In 1922, however, he was assassinated while riding in a car. Villa's name lived on as a symbol of heroism in the *corridos* sung by Mexicans and Mexican Americans.

The forces of rebel leader Pancho Villa ride triumphantly through Ciudad Juárez in 1911.

Conditions in the countryside became even worse as the fighting continued. Pedro Martínez, a villager, recalled:

It reached the point where martial law was declared. There was no way of getting out now. At the end of 1913, and into 1914, you couldn't even step out of the village because if the government came and found you walking, they killed you.

The first village to be burned was Santa Maria, in 1913.... It was entirely destroyed. The [soldiers] had burned everything. The dead were hanging from the trees. It was a massacre! Cows, oxen, pigs and dogs had been killed and the people, poor things, went about picking up rotten meat to eat. All the corn and beans were burned. It was a terrible pity.

During the revolution, Ernesto Galarza and his family made the decision to go north to the United States. Young Ernesto saw the fighting from both sides and encountered first the arrival of a revolutionary army.

The fighters in huaraches and peasant clothes sometimes traveled in the revolution with their wives. The women cooked for their soldier husbands, carried their rifles, toted the bedrolls, and cared for the children as they moved here and there. Whenever the troops camped near the station the women scattered through the neighborhood, stopping at the cottages to ask for a stick of pine kindling to start their fires or a pot of water. When they knocked on our door, Doña Henriqueta invited them to help themselves at the well. She always offered a pinch of salt or a pepper or a small lump of corn dough for tortillas. Behind our closed door, after we barred it at night, my mother explained the ever changing scenes of the revolution that passed along our street. The *soldaderas*, as they called them, often sat outside our door, nursing their babies, the way the women did in Jalco.

The Galarza family traveled by train to Mazatlán, where they stayed for some time, waiting for Ernesto's Uncle Gustavo. The town was under the control of the government forces of President Díaz. Soon, however, a revolutionary army arrived, placing the town under siege.

One night we heard rifle fire from the direction of Casa Redonda. José [another uncle] went out for news. He came back to tell us that the *revolucionarios* were moving in from the north along the railway line and that all the roads leading out of the city had been cut. We were besieged.

The siege lasted many weeks. There was gunfire in some part of the town every night. Shoppers going to and from the market walked close to the walls of the houses and ran across the streets. Food and water became scarce. I scurried across the courtyard on errands to exchange food with our neighbors....

The safest corner in our apartment was in the kitchen.... Doña Henriqueta and I sat there throughout the night. Crouching in the dark I heard the rifle shots of the revolution, something like 'whee-oo,' swift and short, like the snap of a *chicote* [cigar].

For me it was not a night of terror, since I did not understand the danger in each of those musical zings....

We knew almost before anyone else on our street that the siege was over. José, on one of his job hunts, had seen the advance units of the revolution by the cemetery and had found out that they would take the city in force the next morning. Everyone in the *barrio* [neighborhood] came out to watch the entry of the victorious troops. They went by on foot and on horseback, old men and young men in work clothes, shoes or huaraches or boots or in bare feet. Some wore bandoliers and others ammunition belts, carrying their rifles this way and that on their shoulders or resting them across their saddles. Many wore pieces of green-white-and-red ribbons stuck in their hats. We joined the mob and followed it to the plaza where a revolutionary general spoke from a balcony. At the end of the speech the men raised their rifles and fired into the air, yelling, 'Viva Don Francisco Madero' and a lot of other *viva's* I couldn't make out. Standing in the doorway of a shop facing the plaza we cheered as the revolution took possession of Mazatlán.

Josefina Aguilera recalled her family's trip north in 1916, when refugees from the fighting were scattered throughout the country.

Although we were all poor, we brought along a large quantity of food which our relatives had provided so that we would have enough to eat. While we were waiting [at a railroad station], many children and women approached us, begging. We were forced to give them some of our supplies.

"Don't give them food," said my uncle, "because they will end up with it all, and then nobody will give you anything to eat."

But the women in their shawls, who were just skin and bones, were pitiful to see, so I gave them what I could. Soon we had no more. We had no house nor anywhere else to go. There was a large crowd of people waiting just like we were....

Finally the train came and everyone ran to get into a coach. We didn't succeed and climbed on to a car. There were people everywhere.

Who knew where we were going? The fact is that we ended up on top of an oil tanker.

Finally, we came to Torreon. We got down with our tattered clothing stained with oil. We stopped there because the railroad track was torn up. Soldiers and other poor people were all around. We had nothing to eat. I was with my baby; my aunt with hers and her two other children. All we could do was cry. Some vendors watched us from their stalls. A few gave us a sip of coffee, a taco, or something. There was a great deal of suffering....

[When the track was fixed, people] prepared their bags in order to be ready. We all wanted to get into a coach so that we

Women like this soldadera fought alongside the men in the revolution.

When the armies of the revolution passed through villages, they frequently gathered up young men as recruits. This young man, however, poses as if he were proud to be fighting for the ideals that inspired some of the revolutionary leaders.

The railroad built to connect central Mexico and Victoria, Texas, 600 miles apart, was intended to help the northern part of Mexico develop factories and mines. Ordinary people viewed it with fear, for the young men who boarded it in search of work often did not return—as a verse from a corrido *called "Iron Road" describes:*

Listen, Listen,
Hear her roar down the track,
She's coming for a load of men
That she won't be bringing back.

Throughout Mexico, revolutionary armies seized control of the railroads that would carry their troops into battle. Refugees fleeing the fighting often had to give up their places in railroad cars to the soldiers, and many simply climbed on top for the ride.

might sit in a chair. Those that won, won; those that didn't—well, once more we were on top of a car.

I carried my baby in my shawl so that it couldn't fall out. My sister sat on one side of me and my husband on the other. We clasped each other very tightly.

From time to time, the train jerked. Suitcases, hats and children went flying from it. The mothers wanted to fall off too, but those who were holding them from above stopped them. How could the babies who fell survive? We cried and prayed.

"Oh, Jesus, Mary and Joseph! May God be with us!" Our suffering was incredible, but finally we arrived in Kansas.

Lucia Martínez never forgot the horrors she saw when she and her family fled Mexico for the United States.

On the twenty-fifth of June, 1916, we left Aguascalientes, Mexico. Our passenger train was stopped and we were told that we couldn't stay in the coaches but had to climb on top of the roofs of the cars. Since it was during the Revolution, soldiers took the coach seats whenever they wanted them....

Along the entire route from Aguascalientes to Zacatecas, we saw the bodies of many men hanging from posts. They moved with the air currents. Some were soldiers of Villa, others were Carranzistas. If the Carranzistas were the first to arrive in a town and found men whom they thought were Villistas, they hung them. The same was true of the Villistas.

The sight didn't astound anyone, because we all knew that the revolutionaries killed people wherever they found them. The men looked dreadful with their tongues hanging out, but no one paid any attention because it was the same thing everywhere.

Since the soldiers didn't leave much food for the ordinary people, my mother had brought along a large sack of bread. It was all that we had to eat. The others had nothing—nothing at all. When we got hungry, we asked her for something to eat. She took a cup and filled it because now the fat little loaves were in pieces. As soon as the women and children saw the food, they all began to ask for some too. In a moment the whole sack of bread was gone. From there to El Paso we had nothing to eat. They didn't sell anything on the train. Everyone had to bear up. The only thing we had was water.

Throughout the 20th century, poverty has continued to bring Mexicans to the United States. In the 1950s, a Mexican immigrant explained:

Down in Mexico....they are still very poor. There are lots and lots of farmers who are still paying their workers maybe 8 pesos a day. That's about 65 cents.... The results—well, you try it. Let's say you have a wife and five or six kids, and probably your father and mother to support.... Well, on that kind of money you will probably be able to have only one meal a day. That is very common in Mexico.

In 1988, Rosa María Urbina told an interviewer why she decided to leave Mexico for El Paso, Texas.

When I was a teenager, I worked as a hairdresser in a beauty salon, cutting hair. My first husband was a mechanic, fixing cars.... When he died in 1984...my children were nine, seven, and three years old. I had to find a way to pay rent and feed them.

At the time, the economy in Mexico had become horrible. Inflation was going crazy. The peso jumped to 500 per dollar.... I found a job working on an assembly line at a factory.... I would go into work at 4:30 in the afternoon and stay until 2:00 A.M. I was paid only 7,000 pesos [$14] each week. That was not enough to feed my kids. And I didn't have any relatives or friends to watch the kids while I worked. So I had no other choice but to put them in a special institution, like an orphanage.... This upset me very much. But with my husband dead, and no other form of support, there was nothing I could do.

My only hope was to cross the river to the United States. If I could find a job that paid enough money, my children could join me. I wanted them to have an education and a proper life...to be someone.

Luis Murillo had fought as a revolutionary, but his early enthusiasm turned to disillusionment. He explained why he left Mexico:

I looked for a long time but everything had stopped, factories, mills, everybody was without work. With the farms burned there weren't even any tortillas to eat, nothing but maguey leaves....

The condition in which the country is now is nothing more than a killing of brothers, one by the other and now they don't even know why.... I said to myself that it didn't look like a government or a fight over convictions. Why should I fight now?

As the armies of the revolution roamed through the country, families were driven off their farms. Refugees like these tried desperately to make their way north to safety in the United States.

After scaling the high wall that today runs along parts of the U.S.–Mexican border, a young Mexican tries to run away from the migra, or Border Patrol officer.

CHAPTER THREE

BACK AND FORTH

The history of Mexican immigration, both legal and illegal, has been a back-and-forth process. At times, the U.S. government has officially encouraged Mexicans to enter this country. But the United States has also adopted policies to block immigrants or even to force Mexicans who lived here back across the border.

There has been no Ellis Island or other main entry station for Mexicans. Instead, Mexicans have made their way north to one of the twin cities that dot the border. From the Mexican side they crossed into the United States, sometimes after obtaining an official visa at a processing station but often not bothering to do so. The major entry places were through Tijuana, Mexico, into San Diego, California; through Nogales, Mexico, into Nogales, Arizona; through Ciudad Juárez, Mexico, into El Paso, Texas; through Nuevo Laredo, Mexico, into Laredo, Texas; and through Matamoros, Mexico, into Brownsville, Texas.

Each pair of twin cities along the border is virtually a single economic unit. At one time, the border actually ran through a store in Nogales. The proprietor of the store evaded the tax systems of both countries by stepping to one side or the other when making sales.

Today, the economies of the border cities are still linked. Every morning, people leave their homes in one country and commute across the bridges to work in another.

The most important factor in Mexican immigration is that there is no ocean to cross. Barbed-wire and chain-link fences have not kept Mexicans from walking across the border nor have radar and night-vision binoculars. So no one really knows how many Mexican immigrants have arrived in this country, even in recent years.

The greatest number of recorded immigrants came in the 20 years after 1910, when the Mexican Revolution erupted. Officially, the U.S. government counted almost 700,000 Mexican immigrants between 1910 and 1930, but the true figure is probably far higher.

During the 1920s, some new Mexican arrivals began settling farther north. The Mexican population of Chicago, for example, increased from 4,000 to 20,000 between 1920 and 1929. Throughout the upper midwestern states of Michigan, Indiana, Illinois, and Ohio, Mexicans found work in the booming new automobile industry and in related industries such as tires and steel.

Mexicans made up the greatest number of new immigrants during the 1920s because in that decade the U.S. Congress passed a series of laws intended to restrict immigration from Europe and Asia. Laws passed in 1921 and 1924 set quotas for immigrants from European countries but not for those from the Western Hemisphere.

However, the antiforeign spirit of the time led to the creation of the Border Patrol in 1924. It was assigned to patrol the 2,000-mile-long border between the United States and Mexico—an impossible task for a force that at the time numbered only a few dozen men.

In theory, a Mexican entering the United States had to pay a $10 fee to obtain a visa from a consulate in one of the Mexican border cities. Instead of wasting money on this bureaucratic process, Mexicans continued to slip across without official papers. Because some simply waded across the Rio Grande, Mexican immigrants became known as "wetbacks."

Until the end of the 1920s, despite the creation of the Border Patrol, there was little concern in the United States about Mexican immigration. In Texas, California, and the southwestern states, Mexicans provided cheap labor for farms, ranches, and factories. Many took up permanent residence here, raising families and building a new life. By 1930, 70 percent of all Mexican Americans lived in either Texas or California. As early as 1925, more people of Mexican descent lived in Los Angeles, California, than in any Mexican city except the capital.

Then, in 1929, the Great Depression began. Millions of people were thrown out of work. Racial prejudice made Mexican Americans a target of the government's policies. One solution to widespread unemployment was to reduce the number of workers. According to this way of thinking, Mexican Americans held jobs that "real" Americans should have.

Many local governments began repatriation programs that sent hundreds of thousands of Mexicans, both legal and illegal workers, back to Mexico. The federal government also launched deportation drives, although these affected a smaller number of people than the local campaigns. Among those expelled during these repatriation programs were many children and young adults who were born in the United States and were therefore legally citizens.

In 1941, however, the United States entered World War II. With millions of men serving in the armed forces (among them, more than 300,000 Mexican Americans), labor shortages once more arose at home. The governments of the United States and Mexico reached an agreement to encourage Mexican *braceros,* or laborers, to come to the United States. Braceros (the Spanish word is derived from *brazo,* meaning "strong arm") were recruited in Mexico and transported across the border in buses or trains. On arrival, they signed contracts with the owners of large U.S. farms.

The terms of the bracero program stipulated that the Mexican

Mexican families cross the Rio Grande near El Paso, Texas, during the Mexican Revolution.

laborers were to receive a minimum wage (then 50 cents an hour). Their employers would provide proper housing, food, and health-care facilities, and the Mexicans would not be subject to discrimination. These agreements were violated so openly in Texas that the Mexican government forbade the braceros to go there. Texan farmers merely turned again to the illegal immigrants, who were once more coming across the border in large numbers.

The bracero program did not end with World War II. It proved too profitable for U.S. farmers to have a cheap source of labor, and braceros continued to be imported until 1964, when the United States finally ended the agreement.

Some Mexicans applied for the bracero program but were turned down. They responded by joining the swarm of illegal immigrants who crossed the border. The illegals did not receive the same wages and other benefits that the braceros did, and so they were even more desirable from the standpoint of employers in the United States. Between 1954 and 1958, the U.S. government carried out Operation Wetback to identify and expel illegal Mexican laborers. About 3.8 million of them were returned to Mexico.

In 1965, a year after the bracero program ended, the United States passed the Immigration and Nationality Act, which set an annual limit of 120,000 immigrants from the Western Hemisphere. Eleven years later, Congress set a ceiling of 20,000 immigrants from any one country in the Americas. This had little effect on the influx of illegal migrant laborers from Mexico. The trip to the United States had become a way of life.

In 1986, Congress passed a law that granted temporary resident status to all illegal immigrants who had arrived before 1982. In addition, the law granted that same status to any agricultural laborer who had worked more than 90 days in this country during 1984, 1985, and 1986. This proved a boon to some Mexicans, who now were eligible for social services and various benefits that employers are required to provide.

The 1986 law also provided penalties for employers who hired illegal immigrants. It was hoped that these penalties, along with tougher enforcement of previous laws, would discourage future illegal immigration.

That hope has proved to be a vain one. Though the Border Patrol uses spotlights, radar, and barbed-wire fences along the border, uncounted numbers of illegal immigrants still head north from Mexico every year.

As the Border Patrol grew tougher in capturing and turning back illegal immigrants, Mexicans began to rely on guides known as "coyotes." These men or women are experienced at evading the *migra,* as Mexicans call the Border Patrol. For this service, of course, the coyotes charge high fees, and immigrants often lose what little savings they have as "the price of admission." The guarantees offered by coyotes are sometimes worthless promises, and Mexicans have died inside steaming hot, airless trucks provided by coyotes. In addition, those crossing the border illegally are easy prey for hoodlums who know they can rob, rape, or kill the defenseless immigrants.

The attraction of the United States for Mexicans is easy to understand. The average income of Mexicans, in their home country, is about $2,000 a year. A laborer in the United States, receiving the minimum wage of $4.25 an hour, can earn four times that. Of course, the Mexican immigrant soon finds that food, clothing, housing, and everything else costs much more in the United States than in Mexico.

José Varela kept this photograph, taken on the day in 1919 when he and his family crossed the border into the United States.

Many families are among the migrant laborers who cross the border every year to plant and harvest the crops. Some men also come here to make money to provide for their wives and children in Mexico. Today, however, most Mexicans travel to urban areas, and women are an ever-increasing percentage of the migrants. The term *immigrant* really does not apply to many Mexicans who cross the border for temporary jobs and then return. Such people are better termed migrant laborers.

Many others, though, have come to stay. The census of 1990 reported about 13 million U.S. residents of Mexican descent, about half of whom live in the western states. The number of Mexican Americans grew by about 5 million people between 1980 and 1990—an increase of more than 60 percent.

But there are millions more Mexican Americans who have arrived illegally and avoid the census takers. In the 1980s, about 60 percent of all illegal immigrants to the United States came from Mexico. The San Diego office of the Border Patrol captured and returned 427,000 illegals in the year 1985 alone. (Of course, many included in this figure were the same people, making try after try to enter.)

Critics of the immigration policy say that neither the United States nor Mexico really wants the continual back-and-forth immigration to stop. U.S. agribusinesses depend heavily on Mexican labor, and the cost of the food on every table in the United States would soar if Mexicans were not available to plant and harvest it. For Mexico, plagued by unemployment, the emigration of its citizens is a social safety valve. And the money that Mexicans bring or send home from the United States is important to Mexico's economy.

As long as the great disparity in wealth between the two countries exists, the United States will continue to be a magnet for Mexicans. Today a "green card" work permit is a ticket to survival.

COYOTES

There are "coyotes" all along the border—people who guide or transport Mexicans across for a price. Jesús Manuel Hernández Tovar is one of them. He lives in Ciudad Juárez, just across the border from El Paso, Texas. Hernández does not like the term coyote; *he prefers* pasadore, *or "guide." He explained to an interviewer his methods of "passing" people across the border.*

This is the easiest place to cross. Over there, farther west, it is easier because there is no fence and the river isn't as high, but it is very dangerous. There are dangerous people there, *narcotraficantes, drug adictos, y rateros* [drug smugglers, addicts, and robbers who prey on illegal immigrants]. So it is easier to cross here. We pass people on rubber rafts or inner tubes because sometimes there are people who don't know how to swim. Sometimes we have to pass babies and pregnant women and elderly women. The river isn't very high now, but at times it is very high and fast. It can be very dangerous....

Here we see people coming from all over Mexico, from the south, and even from Central America...but most of the people who cross are from Juárez and they work or have family in El Paso, so they cross back and forth every day. We have customers that cross with us every day because they know us and have confidence. They are our steady customers. They cross with us in the morning and come back over the bridge at night.

In 1977, Miguel Torres crossed into the United States four times. He told an interviewer how a coyote helped him.

I went to Tijuana.... There's a person there that will get in contact with you. They call him the Coyote. He walks around town, and if he sees someone wandering around alone, he says, "Hello, do you have relatives in the United States?" And if you say yes, he says he can arrange it through a friend. It costs $250 or $300.

The Coyote rounded up me and five other guys, and then he got in contact with a guide to take us across the border. We had to go through the hills and the desert, and we had to swim through a river. I was a little scared. Then we come to a highway and a man was there with a van, pretending to fix his motor. Our guide said hello, and the man jumped into the car and we ran and jumped in, too. He began to drive down the highway fast and we knew that we were safe in the United States....

[After two months, the immigration authorities caught Torres and took him across the border to Mexicali.]

When we arrived in Mexicali, they let us go. We caught a

An inner tube serves as a temporary boat to take Mexicans into the United States. The concrete ramp alongside the river is decorated with slogans like "Wetbacks," "Ilegales," and simply "¡No!"

Graffiti covers this post, one of many erected by surveyors in the 19th century to mark the border between the United States and Mexico.

bus to Tijuana, and then at Tijuana, that night, we found the Coyote again and we paid him and we came back the next day. I had to pay $250 again, but this time he knew me and he let me pay $30 then and $30 each week....

Altogether I've been caught three times this year and made the trip over here four times. It's cost me one thousand dollars but it's still better than what I was making in Mexico City.

After Lupe Macías's husband immigrated to Los Angeles in the 1980s, he encouraged her to join him. She went to Tijuana but was afraid to cross the border by herself. Finally, as she recalled later, she was introduced to a coyota, *a female coyote.*

She gave me identification and said, "Look, this identification is yours. You are going to pass through the border, but you can't be at all nervous. No shaking! No dancing! You have to pass as if it were nothing. Just in case they ask you, you will tell them you are going for an errand in San Ysidro.... Listen, you have two daughters. One's name is Monica, she is thirteen, and your other daughter is six."

I said, "Do you think I am going to remember this?"

"Yes. Now tell me, what is your name?" I was supposed to tell her in English. Well, I forgot. She said, "No, no, don't be nervous."

Aye Dios, I said, "This isn't going to be possible."

"Come on, let's see. What is your name? How many children do you have?" I had to learn this in English. *Dios mío,* I was perspiring with fear. "Listen," she said, "you are going to be carrying a bag because you have an errand. You and your husband work in San Ysidro."

Well, I learned everything and just as soon as I did she said, "Okay, you're ready, let's go." We went right to the line, you know where the people walk across. It was a long line with every sort of person carrying all kinds of things. Many were gringos carrying bags and bags of souvenirs and things they had bought in Mexico. There were a lot of Mexicans too. Well the señora told me, "You get in line here and I will go ahead. In the case that you pass, I will be waiting and give you a sign. Then follow me, but a little behind." I wasn't going to talk to her or even look like I knew her.

The line was very slow. After awhile, I could see where they were checking everyone. Oh, no! (The border guard) was a very fat woman—like this—with her *pistolas. Aye madre mía!* And she was sending back so many people. I went asking the benediction of the little saints. Aye, please don't let her turn out the lights! Well, I just kept in the line and as I came up to where they were checking people there was some kind of disagreement or something. A man was in front of them with his bag open and they had everything out, they were speaking pure English, all of them, and well the fat woman...just glanced at my card and went back to talking with the other *migra* and off I went. Aye little *Dios,* I passed, I passed!

A group of would-be immigrants waits in a Mexican village for the coyote who will guide them across the border.

EVADING THE MIGRA

Late in her life, Isabella Mendoza recalled how she evaded the border guards when she was two years old.

I was born in Mexico in 1913, and when I was four months old, my father took off for the United States. At first he sent money for us and then he got lost. I don't know; he got lost in the United States. He lost himself somewhere. And my mother said, "I'm going up there to look for him. I'm going to find him." She got a little money together—sold things in the market—and she came up here with me and her sister; that's my aunt. We went to the border and came across. We didn't have no papers, no nothing...but in those days it wasn't so hard. My aunt knew someone who knew a guard at the bridge, and she went up and started to talk to the guard, and she gave him eight dollars. And while they were talking, my mother took me on her shoulders and waded through the river on the other side. It wasn't so deep then; the water just came up to her shoulders. And when we got up on the other side, my aunt stopped talking to the guard and he let her go across because she'd paid him the eight dollars.

Today's Border Patrol, whose agents are known to the Mexican illegal immigrants as the migra, *is equipped with modern technology—yet it still fails to stop countless numbers of determined Mexicans who are armed only with dogged persistence. Ramón Gonzales was one of the migrant laborers who frequently crossed the border illegally. In his autobiography, written in 1966, when he was 45, Gonzales described the experience.*

We never jumped the border in the daytime. Every time it was in the night. You can't see, you know. You can't see, you go just on direction. And you figure, Los Angeles is in this direction, and you cross a mountain and come to another one. We would sleep out in the bushes. You have a jacket or something and you just lay down there in the bushes. You bring food for a certain number of days.... So you sleep under a tree there all day, until when it gets dark you start going again. That's how we used to do it....

One time I crossed the border from Tijuana to go to Los Angeles. I had to walk from Tijuana clear to Los Angeles. I crossed the mountains; in the mountains we were very high, you know. I had to take about fourteen days to get to L.A. by crossing the mountains, because if you go by highway there would be a lot of Immigration officers.

This boy went with me, and he couldn't make it through

Immigrants scale the steel wall that separates Tijuana, Mexico, from the United States. The wall is made from mats that were left over from building aircraft runways during the Vietnam War.

the mountains, crossing the high mountains. You know how tired you get climbing, going down on another mountain, up and down. And this boy, he got so sick I thought he was going to die on me. I thought this boy was dying, and I didn't know what to do. He fainted from walking too much, and then I had to put water on him to bring him back, and then we rested for half a day. Pretty soon he was all right....

Then another time that we crossed from Mexicali we walked about fifty miles. Then we got a train, and these trains pass through a lot of towns where the Immigration inspect the train, you know. We stopped at one town in California...for them to inspect the train. When the train stopped there we had to get off, and everybody started running because there would be a lot of Immigration. When I seen the train slow down, I would jump from the train and hide someplace. Then when the train would start up again, we would hike the train again.

There were a lot of people that used to get on those box-cars where they put the ice. Then they would close the door, and sometimes they would jam the door, and then you can't open it. A lot of wetbacks died. They starved because they couldn't open the door. And a lot of those tanks, you know those oil tanks, a lot of people used to get into them, and they would suffocate. That's why I never used to get into those places. Sometimes we used to jump off the train, and we would lay around in the orchards. All we had for food was oranges and grapefruits. We used to have it rough, sometimes, jumping the border into the United Sates. I haven't tried it no more.

After some illegal immigrants were killed while rushing into traffic to evade the migra, *California has found it necessary to erect road signs warning drivers.*

A group of Mexicans waits to cross the border. When the migra *patrol van in the background has passed by, they will have an opportunity to jump the fence.*

A FOOT IN TWO COUNTRIES

Many Mexican Americans retain a strong tie to relatives and friends in their homeland and often return for visits. David F. Gomez, a priest and Chicano activist, recalled trips back to Mexico with his parents after the family settled in Los Angeles in the 1940s.

My parents kept close contact with the land of their birth. It is about 140 miles from Los Angeles to Tijuana, and about 200 miles to Mexicali. Many of my relatives would head south across the border to have dental or medical work done. They went not only because the work could be done less expensively but also because they would be better able to communicate with the doctor in their own language. My parents did not have any language problems, but we would head south anyway just for the recreation and the opportunity to visit with friends and relatives there. My brother's *padrinos* [godparents] lived there, and that was a special relationship that linked our families closely together. Each summer my sister and I would swim in the arroyito [stream] with local Mexican children and enjoy ourselves.

Sometimes the trip back to Mexico was not voluntary. During the Great Depression, Mexican workers were no longer welcome in the United States. Many Mexicans, even those who had become U.S. citizens, were forcibly repatriated to Mexico. During the early 1930s, some 300,000 people were sent back. Jorge Acevedo, one of them, describes the arrival of a van in the early morning to Maravilla, a barrio in Los Angeles.

Families were not asked what they would like to take along, or told what they needed...or even where they were going. "Get in the truck".... Families were separated.... They pushed most of my family in one van, and somehow in all the shouting and pushing I was separated and got stuck in another van. It was a very big one with boards across it for us to sit on. Nobody knew what was happening or where we were going. Someone said, to a health station.

We drove all day. The driver wouldn't stop for bathroom nor food nor water. The driver was drinking and became happier as he went along.... It was dark when he finally ran the truck off the road. Everyone knew by now that we had been deported. Nobody knew why, but there was a lot of hatred and anger.

In some places, the border is marked only by a plastic curtain. These immigrants have already pushed through it and are waiting only for the photographer to leave before crossing into the United States.

Joaquin Avila, Jr., whose parents moved to Los Angeles before his birth, recalled that they frequently returned to their hometown in Chihuahua in the 1950s. Young Joaquin found that the other children did not regard him as a Mexican.

When I was growing up, I spent a lot of time in Chihuahua, summers. I had a good time. My grandfather was a watchmaker. He had a little shop in his house. When I would go there I would get to watch him, and he would give me a couple of watches and little spectacles and little tools to tinker [with]. When my cousins in Chihuahua used to get mad at me, they'd call me *pocho* [a term of disapproval meaning someone who favored the culture of the United States rather than that of Mexico]. I wouldn't call it discomfort when we go back to Chihuahua, but there is a difference, a perception that we're pochos, almost like half-breeds living out here.

Many Mexican Americans are at home in both cultures. Sergio Rodriquez, who is a mechanic in Houston, goes home to Monterrey, about 500 miles away, every weekend.

I am making my life here, but I go back to Mexico a lot because I left part of my life there. Here and there, I feel at home in both places, and I go back and forth because it is not difficult. Lots of people do it.

In 1993, Julio Guerrero, a Houston welder, had been in the United States for eight years. But he is a third-generation Mexican American. His grandfather followed the crops as a seasonal farm worker in California and the Southwest. Guerrero's father worked in factory jobs in Houston but retired to Mexico. Though Julio himself is a naturalized citizen, his wife and children live in Monterrey, Mexico. Each weekend, he commutes.

I would fight and die in the American army if they called me. That's how much I love this country. But as a person, I feel Mexican, and I want my children to feel that way too, even though I hope they will come to live here when they are older. For now, they come here to visit, but they'll stay in Mexico because it's easier, safer to raise them there.

A boy named Tito immigrated to Los Angeles from Tijuana, Mexico, in the early 1980s, when he was four years old. As a teenager, he described the importance to his family of returning to Mexico regularly.

Every year my family visits the town of Guadalajara where my mother comes from. For me it's like going to a paradise. I am breathing fresh air. The people are friendly. It's not like big cities where it's so crowded that everybody has their own little circle around them. "Don't get too close!" they say with their eyes. In cities there are so many people around I don't know who's going to do what to who.

In my mother's hometown, you can say "Hi" to someone and they'll smile a big fat smile and say "Hi" back. Then it's

These Mexicans sought to enter the United States legally in 1938. They are waiting at the U.S. Immigration Station in Ciudad Juárez for passes that will allow them to go to El Paso, Texas.

Susana Hurtarte was born in Tijuana, Mexico, one of the border towns. Her father owned an advertising and public relations firm. Both parents were involved in charitable work in their community. But as Susana recalled, Tijuana suffered from its location near San Diego, California:

Being born in Tijuana can be an unusual experience. The city was founded by a woman who became wealthy selling liquor to American servicemen stationed in San Diego, California. During Prohibition men would cross the Mexican border and buy alcohol at local bars. As the town grew, two distinct and separate societies emerged: one that was non-drug-related, and [one that] survived entirely on drug-related activities. Although I was brought up as *una niña appropriada* [a proper child], I saw persons in the city who existed totally under the influence of drugs, and consequently I could never be enticed to use them.

"Don't I know you? Aren't you Carlos's grandson? Yes? Oh, we're great friends. We used to go to school together."

My mother moved from that town to Tijuana when her father lost his bakery business. She was very brave and strong. She found a waitress job and sent her parents money. My pop came into the restaurant once and really liked her. He went there again and again and she started giving him special attention. Then they got married and had us three kids. For many years she didn't want to move to the United States because she didn't think it was a good place to raise us. Finally, she came and has had to struggle here, too, but she's always survived and done okay.

In 1965, the Mexican government began the maquiladora *program to encourage U.S. industries to build factories on the Mexican side of the border. Using low-wage Mexican labor, the* maquiladoras *(assembly plants) produced such goods as electronic products, clothing, and shoes. Cesar Caballero, director of the University of Texas at El Paso library, commented on the economic benefits that the* maquiladora *program produced on both sides of the border.*

A lot of our neighbors are Americans who have come from the North to manage the *maquiladora* industries that are being set up across the border in Juárez.... For every twelve *maquila* jobs created in Juárez, one job is created in El Paso....

El Paso has a strong business tradition with people in Mexico. If the Mexican economy stabilizes, it will create a golden opportunity for small-business people here. Many Mexicans come to El Paso to shop and trade, but the past few years there has been a downswing because of their economic problems. The popular feeling is that the *maquiladora* program is here to stay.

People who live in one of the twin cities along the border sometimes celebrate their bicultural heritage. In 1942, a group of Mexicans watches a parade at a fiesta organized by the schoolchildren of Matamoros, Mexico, and Brownsville, Texas.

Dr. Celestino Fernández, an administrator and professor of mathematics at the University of Arizona, still returns frequently to the village in Mexico where he was born. He explains what it is like to have a foot on each side of the border.

I feel Mexican and I behave American. Inside, my feelings, my values, my attitudes, my beliefs are based in Mexican culture, but my behavior is very American. I feel very comfortable here. I understand the system and I can work it. My wife says when I'm in Mexico I'm Mexican. I know that system as well and can fit in and behave Mexican. I'm the only one in my family who is a naturalized American citizen.

I haven't left Mexico behind. I always carry a letter in my briefcase from my grandmother, written four years ago when I was promoted to Associate Vice President for Academic Affairs. She knows that I work at a university. She doesn't know exactly what I do, but she knows that I have an important position. She wrote me a letter of congratulations. She says, "We know that you've done a lot in terms of education and your career, but you never make us feel uncomfortable. When you come here you're Tino. You treat us as if we were equals." I know exactly what she was talking about because I have cousins who, after being in the United States for five or ten years, will go back to Santa Inez and dress in a tie and suit—in this little town. It makes no sense whatsoever, right? It's presumptuous, it's irritating to people, and rightly so. When I go there I help milk the cows because I enjoy it. I enjoy riding horses in the country. There are different chances, different opportunities.

Mexican culture has a certain wisdom I appreciate. My grandfather is *Don* Chema. You don't just call yourself by the title of *Don*. I can't be *Don;* I'm not old enough, I'm not wise enough. Unlike American culture, Mexicans value age. There is respect for someone for simply being older. They don't feel bad if they are ninety, because they are still respected.

Border cities such as El Paso, Texas, and Ciudad Juárez, Mexico, are so close that some people live in one place and commute to work in the other. Ricardo Anguilar Melantzón is one of them. A writer, he teaches at the University of Texas at El Paso and lives with his wife and daughter in Ciudad Juárez. He commented on the experience of crossing the border each day.

There are words here on the border that can't be translated into either language. It's a very, very interesting culture, especially if you constantly go back and forth....

The border has been part of my life and my wife Rosita's life, our daughters too. My father-in-law was kind enough to give us this house as a wedding gift. So if we would leave Juárez to live in El Paso we would have to leave this house. It's nice and I'm comfortable here. It's right downtown. For example, I can just walk downtown and buy reviews that some people have to wait months to get in the library....

Victor Villaseñor

"And there came Roberto's mother on the other side of the river. Crying, and shouting, and praying, and bellowing like a cow with calf. She loved him. And she wanted him to stay home. To please, for the mercy of God, not go. Because he'd never return. She just knew it. And for two kilometers she followed on the other side of the moon-bright river, crying her love and fear and heartbreaking sorrow. Roberto began to cry and told her to please go home. But still she came, rosary in hand, bellowing her love." These lines are from *Macho!,* the classic novel of Mexican immigration by Victor Villaseñor.

Villaseñor was born on May 11, 1940, in Carlsbad, California, the son of Mexican immigrants. He had trouble at school because he was Spanish-speaking and had a learning disability that made reading difficult. Discouraged, he dropped out of school and went to work in the fields and as a construction worker.

A trip to Mexico brought Villaseñor face to face with his own identity and culture. He took a new pride in his heritage; he started to read widely and began to teach himself to write fiction. After Villaseñor returned to California, he continued to write in his free time from construction work. All together, he completed 9 novels and 65 short stories. It was not until 1973, when *Macho!* was published, that his literary career was on its way. In 1983, he wrote the television screenplay for "The Ballad of Gregorio Cortez." Starring Edward James Olmos, it was based on the writings of the Mexican American Américo Paredes of Texas.

Villaseñor spent years of research on *Rain of Gold,* his multigenerational family epic of the Mexican American experience. He drew on his personal experiences with his extended family and conducted interviews with relatives. Villaseñor uses traditional elements of Mexican folktales and the oral tradition to tell his story. Published in 1991, *Rain of Gold* has brought to millions of readers the hardships and struggles and the triumphs of the Mexican American experience.

Mexican American families living in the United States frequently return to Mexico to visit relatives. This family in El Paso is returning after a day's shopping in Ciudad Juárez, in 1937.

I was born in El Paso, September 16, 1947.... My dad's family immigrated from Zacatecas at the end of the last century.

My dad worked for the electric company in Juárez; he's an electrical engineer. And my mom has always been a homemaker, but she's also dabbled in trying to get some businesses off the ground. She has some apartment houses. We lived here in Juárez. My mother's mother and father lived across in El Paso. So I used to spend a lot of time over there also. You could say that ever since I was a child I have been going back and forth....

I am one of those people, weird people around the world that live on a border and have dual nationality. I don't have a Mexican passport, but according to Mexican law I am a national of Mexico because my dad is Mexican....

I guess we all have problems crossing. They [the guards] know me. All they have to do is punch the number of my [license] plate in the computer and they know who I am: my name, my work, what I've done.... Every once in a while, there's someone there, usually a Chicano...somebody just like me, who stops me and makes my life difficult. One day they wanted to take the tires off my bike. I said, "Why don't you bring in the dogs and have them sniff first?" So they did and it was okay. You know, what can I carry on a bike? Sometimes they just want to make it difficult....

I consider myself a Chicano. I was born over there, I received the adult part of my education in the United States as a Mexican. But I can't separate myself from what I feel for Mexico. It just happens; I think it is a question of geography. If I would have been born and raised in Las Cruces I would have been a totally different person. To be honest with myself, I have to be a Chicano. I can't be a Mexican and not be an American. And I can't be an American without being Mexican.

In the 1980s, a recent immigrant wrote to her family in Mexico. She had left for the United States without her parents' permission.

Hello Papa, Mama, brothers and sisters,
I hope this finds you very well in spite of everything. By now you will have found out I'm in the United States (California) and you will be asking yourselves what I'm doing here.

Before anything I want to ask you to have a lot of trust in me. I came with the desire to work here a few months in order to gather a little money.

I have great faith that it's going to go very well for me, and this is important. Here everything is different from Mexico, but I assure you I'm never going to forget anything I learned from you!

I remember you a lot and this is going to be decisive in my behavior. *I am not going to fail you! Of that I assure you!!*

When I arrived in the U.S. Wednesday morning I had two addresses of people I could head for.... I opted for [friends] and

I'm sure that if you knew them you'd quit worrying. They're Mexican, they have two daughters, little ones.....

I just want to work a few months and return to you to continue in school....

I want you to be calm, my being alone right now is going to help me a lot to start being responsible for myself.

I'm going to write constantly to tell you how it's going with me. I hope you also write me, even if it is to box my ears. Until soon.

Loving you all,
Patricia Zarate

[P.S.] Just now we went to mass at a church called la Asunción de Maria, it was a bilingual mass with mariachi.

The money sent home by Mexican immigrants is often important to those left behind. A mother in Mexico wrote this letter in 1989.

Dear son and daughter-in-law:... It gives me great pleasure that you have written me my letter, I loved the words you put down very much...your grandma was very distressed because she didn't have enough for [her granddaughters' school] uniforms and I told her don't be distressed, tomorrow I'm going to cash the check in order to bring her her centavos, and I went to exchange the 85 dollars and they gave me 209,000, that is, they took out 3 dollars, and on the way back I stopped by to give her the money.... Please take good care of my son and my grandchildren, anyway, and I also ask you another favor, that you look after my son-in-law, the little goat, that is Toño, take good care of him also greet your mama on my behalf as well as your brothers and your brothers-in-law and family.

Jorge: many thanks son for what you sent me, God will help you more and give you strength to work, that's as much as your mother, the "shrill one" says. I don't say good-bye to you but until soon,

Your mother

A woman dozes while waiting for the international streetcar that will take her from El Paso, Texas, across the bridge to Mexico. She may be one of the Mexicans who cross the border daily from their homes to their jobs.

Sandra Cisneros

"I was/am the only daughter and *only* a daughter. Being an only daughter in a family of six sons forced me by circumstances to spend a lot of time by myself because my brothers felt it beneath them to play with a *girl* in public. But that aloneness, that loneliness, was good for a would-be writer–it allowed me time to think and think, to imagine, to read and prepare myself." In these words, Sandra Cisneros described the experience of her childhood. One of today's best-known Mexican American authors, she has written that she always hoped to gain her father's approval through her literature.

Sandra Cisneros was born in Chicago in 1954, the daughter of a Mexican father and Chicana mother. Much of her early childhood was spent in both Chicago and Mexico City. Her father would get homesick and the family would pack up and move to Mexico. Because of this traveling back and forth, Sandra found school difficult. She was shy and did not make friends easily. She recalled that the teachers, who were nuns, seemed to enjoy belittling their students. Sandra found that she retreated into being an observer, a skill that she can use today in her writing.

Cisneros remembered that her father was pleased that she was going to college because he believed that it was a good way to find a husband. Instead, she discovered that she enjoyed writing. After graduating, she attended the Iowa Writers Workshop. Here, she made the decision to write about her own people. "I knew I was a Mexican woman, but I didn't think it had anything to do with why I felt so much imbalance in my life, whereas it had everything to do with it! My race, my gender, my class! That's when I decided I would write about something my classmates couldn't write about."

The hard work of field labor has already made the face of this young migrant worker seem old.

CHAPTER FOUR

GOING TO WORK

Except for those who fled the chaos of the Mexican Revolution, the great majority of Mexicans who came to the United States were attracted by the promise of jobs. Mexico's population has always grown faster than the country's ability to provide work for its citizens. On the other side of the border, wages were higher and jobs plentiful. As one Mexican said, "The dollar was worth more than the peso."

Mexicans, like members of other immigrant groups in the late 19th and early 20th century, filled the need for cheap labor in the United States. The Mexican vaqueros taught their skills to U.S. cowboys. Mexicans who had worked on railroads at home found new work on the rails that stretched across the southwestern United States. Experienced Mexican miners were recruited to help work the silver, gold, copper, and coal mines of the western United States. It was dangerous and back-breaking labor. Miners who survived cave-ins and accidental explosions sometimes contracted lung diseases underground.

The greatest contribution of Mexican American workers has been their work in agriculture. In 1920, when Texas had become the third largest state in the production of corn, a member of the San Antonio Chamber of Commerce said, "We couldn't do it if we didn't have the labor. Yes, sir, we are dependent on the Mexican farm labor supply, and we know it. Mexican farm labor is rapidly proving the making of this state."

However, California was the state that was to benefit most from Mexican American farm workers. In 1901, water from the Colorado River was diverted into the Imperial Valley, a vast area of Southern California that previously had been a desert. Today, the valley is the most productive agricultural area of the United States. Its lettuce, tomatoes, carrots, melons, and other crops appear on dinner tables throughout the country. Mexican Americans have been the mainstay of the valley's work force.

Workers from Mexico are also important to many other areas of the United States. Migrant workers take advantage of the fact that crops of different kinds are planted and harvested at different times of the year. Taking their families along in small trucks outfitted with living quarters, they follow regular routes throughout the growing season. By planning carefully, the workers will arrive in each place just in time for the work to begin. For some, the annual trek takes them as far north as Michigan, where the cherry-picking work ends the season. Indeed, whole families have repeated this northward trek for generations.

The treatment of the migrant laborers has always been abysmal. They often live in temporary camps near the fields. In 1934, after visiting one such camp, observers from the National Labor Relations Board (NLRB) stated: "We found filth, squalor, and entire absence of sanitation, and a crowding of human beings into totally inadequate tents or crude structures built of boards, weeds, and anything that was found at hand to give a pitiful semblance of a home at its worst. Words cannot describe some of the conditions we saw."

Conditions had not changed more than 30 years later, when Secretary of Labor Willard Wirtz observed a migrant worker camp in California. Wirtz said, "I'm glad I hadn't eaten first. I would have vomited."

The most dreadful part of the migrant experience is that children are part of it. Boys and girls as young as seven spent an eight- or ten-hour day in the blazing heat, stooping to pull out weeds or pick carrots, lettuce, and beets.

Few of these children ever received much of an education, dooming them to spend their lives in the fields at unskilled labor. It was not until the 1960s that the U.S. government began to address the conditions of migrant child labor and lack of education—and even today few laws protect them.

Though some immigrants found better jobs in meat processing plants, automobile factories, and other industries, the United States has never lost its dependence on Mexican agricultural labor.

FINDING WORK

In 1875, Jesús Maria Huerta and his wife, Locadia Herredia, left the state of Sonora in Mexico and settled in Tucson, Arizona. Many years later, their son José described the jobs his father took.

My father used to work with the railroads. I don't know how many years, but he worked a long time. And then when they had a strike he lost the job because the union lost the strike. He used to sell fruit and ice cream. Mostly he would deliver to the bars. Lemons, because they used a lot of lemons and oranges too. He did that for quite a while. And then he got a job at the mines. He used to work at Silverbell Mines, underground. Another job he had was out there working that trail from Sabino Canyon to Mt. Lemmon.... Oh yah! The wood yard! He used to sell wood and deliver in a cart by hand, pushing it around the neighborhood. At the time there wasn't any gas and everybody used to use wood.

As an old woman, Maria Petra Alfaro remembered how she and her husband and children fled to the United States in 1919, one step ahead of the soldiers who were chasing her husband.

Our life was hard and sad because we lived in a period of great need. When men could get work, they earned twenty-five cents a day. So, he who worked for six days, earned a peso fifty for the week. There was no work for women; only for men. I suffered a lot in those days; now I don't; now I enjoy life. Since we came to the United States our suffering has ceased. Here there was enough work for the men. They earned little but it was much better than what they had earned in Mexico. Thus, we must recall our former way of life, how we lived and how we suffered. Today I live in contentment, humbly, because of the happiness which I have found in the United States.

Ramón Gonzales's parents took him to the United States in 1921, when he was about six months old. Ramón's father worked as a farm laborer in Southern California. When Ramón was old enough, he began to help support the family.

When I was about nine years old I used to sell newspapers in San Bernardino.... I used to make good money, because I had a good corner where I used to stay and sell them all the time. Lots of times we make fifteen dollars a day, or eight, or five. It depends on the papers....

Another job I used to do to make money was when I was about twelve years old was work in a circus. I used to sell

These men are making adobe bricks in California in the early 20th century. Adobe, a mix of clay and straw that was cut into bricks and dried in the sun, was a common building material in Mexico and parts of the western United States.

A resident of San Antonio, Texas, around 1968. San Antonio is the largest U.S. city with a Mexican American majority.

paletas. Paletas are frozen suckers.... In San Bernardino they have a place where they make frozen suckers. We used to get them there and sell them in the circus, and when the circus wasn't there I sold them at the beach, like Balboa Beach, close to Los Angeles....

Then I used to sell fruit like cantaloupe and watermelon. I went door to door, knocking on the door, "Would you like to buy a cantaloupe or a watermelon?" Sometimes we used to make good money, selling like that, but lots of times you would be ashamed to knock on the door because there would be girls there. You would be kind of ashamed to be selling fruit like that.

One other thing we used to do to make money was pick grapes. We used to pick red grapes and black grapes and put them on trays. Then the sun would dry them and make raisins. This was still in San Bernardino, when I was about ten or eleven. We got paid one cent a tray. Usually we would pick one hundred trays, so we would get one dollar. Sometimes we picked two hundred trays a day. But now I am a better picker, I could make more than that. But for a young boy one hundred trays is a lot of trays!

Though Ramón Gonzales worked for years in the United States—illegally—he was caught and deported. Then in the 1940s, he took part in the bracero program, in which Mexican workers were allowed to enter the United States legally. However, as he recalled in 1966, luck and chicanery were part of the process.

I came to Mexico City because they were going to send some men to the United States as braceros. So I came, and there were thousands of people waiting there to get jobs as braceros. Oh, I thought, I would never get there because there were thousands. So finally one man told me, "You want to go as bracero?" I said, "Yes." "Well, I will sell you a pass to get in." I said, "How much you want for that pass?" "Well," he said, "give me one hundred pesos." So I said OK. So I gave him a hundred pesos for the pass, and I got in to see the doctor for a physical check and everything. Finally we pass all that physical checking, and they said, "Well, a certain day, you all are going to leave." So they told us to be in the railroad station at a certain time. And we all got in the train;... I was the only one who could speak English.

And when the train came, when the train cross the border, they had to get your name and everything. They had your papers, and everything was clear. We crossed the border, and they brought us all the way, on the train, all the way to Fresno, California. There in Fresno there was a lot of ranchers. They would give, say, fifty dollars for fifty guys. And then one rancher asked, "Any of you guys speak English?" So I said, "Yes, a little." "Well, I want you to pick out fifty of the best you think are good workers." So I told him, "OK," so I picked out most of the guys that came from the town I was born in, and we went to work for him.

As Others Saw Them

The Chicana maid was a fixture in many Anglo homes in Texas and California. Working as a domestic was one way a mother could help support her family. Elizabeth Ray Tyson, an Anglo resident of El Paso, recalled:

Owing to the large Mexican majority, almost every Anglo-American family had at least one, sometimes two or three servants.... The maid came in after breakfast and cleaned up the breakfast dishes, and very likely last night's supper dishes as well; did the routine cleaning, washing and ironing, and after the family dinner in the middle of the day, washed dishes again, and then went home to perform similar service in her own home.

Braceros arriving in Stockton, California, in 1943 to harvest sugar beets.

ON THE RANCHO

Many Mexican Americans in Texas worked on ranches. The work was hard and the life was often isolated. Moreover, the worker was dependent on the patrón *who owned the ranch. One such worker explained:*

I worked for my *patrón* for fifteen years. Then one day the *patrón* told me to leave. He didn't need me any more. You see he bought a tractor, and it could do the work better than me. He was tired, too, of our chickens eating on his grain, and he thought our children should stay home from school and work. We weren't needed anymore, you see. So he got rid of us.... No, I do not think the *patrones* really liked us. Many Mexicanos believed that the *gringos* were being good to us, but I believe they used us like work animals. I know many *sembradores* [farm laborers] who had to leave their place. We did not want to believe our *patrones* did not care. We had nothing else. We could not speak English. We had no land and no education. For many of us the *patrón* was our only hope. But we were never close, not truly *compañeros* [close companions]. The *gringos* were only nice so we would work harder and stay with them until they no longer needed us. No, Mexicanos did not really like their *patrones*. And the *patrones* did not really like us. We were not bonded together like *la familia*. If we could have left we wo uld. If the *gringos* didn't need us, they would have sent us away.

Life on the ranches was hard work for women as well. An old woman described her duties on a ranch in southern Texas where she lived from 1915 to 1932.

I woke up first in the house and put some water to boil in a pan for the children's baths. Then I'd put some cold water in the big tub in the bedroom, plus the hot water until it was warm enough for the bath. I'd bathe all the children, one by one, and dress the smaller ones. By then my husband was also up. I'd prepare the breakfast for him and the children. He would leave by five or five-thirty in the morning. I'd then give something for the children to play with and be busy while I prepared the *masa* (dough) for the tortillas. My older daughter helped with grinding the *maíz* in the *metate* (stone bowl). That would take me about an hour or so to do. Then I'd have to start preparing lunch for my husband; if he was working in a land near the house, he would come home for lunch at about noon. He'd stay home until two or three in the afternoon because the sun was too hot at that time for him to work. Well, I was always busy at the house with one thing or the other.

Lulario Gonzales and his son with the herd of goats that they tended on La Mota Ranch in La Salle County, Texas, around 1928.

In the 20th century, Mexican vaqueros are still in demand for their skills with horses and cattle. These Mexicans eat dinner at a ranch in Refugio County, Texas, around 1950.

To get ahead, the whole family worked together at many kinds of jobs. One Mexican daughter described life on a ranch in the 1930s.

My father was a *cuartero* [tenant farmer] in the *rancho*. We all helped him in *la labor* [field work]. The whole family worked, and my mother stayed home. She had her work too. She would prepare the meat, salt and dry it so that we would always have a fresh supply of meat. She canned food, sewed for the whole family, and kept our small house spotless. Then my father bought a little plot in North Town [Texas]. He himself built the house which is still standing. It is not far from here. *Mi padre santo* [my saintly father] never lacked anything. He bought the plot near the *escuelita* [school] so that we children could go to school by foot. While he continued going to the *rancho* to work, my mother raised chickens and turkeys at home. Later on, after many of us were already married, my father decided to become part of crews and moved to another town where he could find more of that kind of jobs. Then he'd work everywhere in these two counties.

In the ranch life of southern Texas, visiting friends and relatives was an important part of social life. One woman remembered:

Mi *comadre* Rosita often visited me. She was frequently coming to our house some week-ends. She'd arrive on a Saturday afternoon, coming by wagon with everyone, all the children plus the couple. They'd stay until Sunday morning and then head back home. Her children greatly enjoyed the *columpios* (swings) that my husband had hung from the trees for our children. All our children played well together. The house was small, but there was room for everyone to stay overnight. The floors would be covered with *pura muchachada* (many young people). Of course, there was always enough food. Who feeds one, feeds ten.

Conditions in out-lying areas were often primitive. This girl draws a cup of water from a faucet that was probably her family's only source of clean water.

The wedding of Eufracio Rodriguez and Eufemia Zambrano on a ranch near Kenedy, Texas, in 1909.

Mexicans worked in the quicksilver mines of New Almaden, California. In 1857 a visitor described its operations:

We enter the car and in a few moments are rumbling along this under-ground railroad, with no sound to break the silence besides the heavy breathing of our human propellors, who, with swarthy visage lighted up by the dim rays of the candles, seem almost ghastly as they bend to their work. The laborers are all Mexicans and have generally served a sort of apprenticeship in the silver mines of Spanish America....

The mechanics, who are mostly Americans, receive full city wages—from five to seven and the laborers from two to three dollars a day. These last are fair specimens of the reckless, improvident Spanish-American race.... It is of little consquence how much or how little they receive.

Mexican miners in Arizona in the early 1900s.

RAILROADS AND MINES

Carlotta Silvas Martin was born in Arizona in 1917. Her father, Miguel López Silvas, had come north from Mexico to work as a miner for the Magma Copper Company. He met Carlotta's mother, Elena Amanda Romero, in the boardinghouse where he lived. In the 1980s, Carlotta remembered her father's 43 years of work in the copper mine.

The miners worked very, very hard; they worked under intolerable conditions. It was very hot underground; it could get up to 140 degrees. There was no air conditioning like there is now. The dynamite blasts and the equipment down there stirred up dust. I remember when he came home he looked so terrible, so exhausted and tired, that his eyes would slip back into his head. A lot of miners were killed in that era. As recently as May 1966, three miners were killed in a dynamite blast caused by a short fuse.... A corrido was written about it.

Almost all of the miners of my dad's era died of silicosis; silica is a very fine glass dust that gets into the miners' lungs. Eventually it digs a hole in the lungs. It's a very wasting disease. There was a lot of anger. They couldn't breathe, they coughed up blood, and they choked on blood. It was a horrible way to go. Many of the wives of the old-timers took care of their husbands. By the time they died, the wives were ready to die from exhaustion.

When my dad retired, he already had the disease. The mining company had a law that you had to have been underground within the last ten years [to collect sick benefits], and my dad had been working above ground in the change rooms where the miners took their baths. The company used that as an excuse not to pay him his silicosis compensation of $2200. My dad didn't know how to speak English, so when the Labor Relations Board met, I went and gave them a piece of my mind. I said, "My dad has silicosis. Now where in the world do you think that he could have gotten it other than underground? He hasn't worked anywhere else." Eventually, he did get his compensation, but they cut off his $17 a month pension because they said that no one could receive two benefits at the same time....

My father died of silicosis in 1964 at the age of seventy-six. He had always been a strong man, and I am sure that he would have lived much longer if it hadn't been for that.

Newly arrived Mexican immigrants often sought out a renganchista, *or work contractor, who knew where jobs were available. When Lucia Martínez and her family arrived in the United States in 1916, Lucia's father found work through such a man. But his experiences were bitter ones, as Lucia Martínez recalled many years later.*

A labor contractor sent us to a small Texas town. On that trip we only had sardines and crackers to eat. Throughout the three months that we were in that town, my father never was paid more than fifty cents at a time. My sixteen-year-old brother, who worked on the railroad tracks, only received twenty-five cents. The bosses kept a list of what people bought to eat. When a check came, they gave us no money because they said that it had all gone for food. We only had potatoes, beans, flour, coffee, and sugar.

One day the commissary agent got mad at my older brother and said he was going to kill him.

"We can't stay here any longer," my father announced. "Any day that man might attack Elidoro. Let's leave."

We didn't have a cent. Absolutely nothing. We left at one in the morning so no one would see us. We packed some suitcases with bedclothes and a little clothing. My mother fixed some biscuits and beans.... We walked along the track for 18 miles until we came to another town. At five in the afternoon we arrived. Some Mexicans who lived in houses made of railroad ties let us stay outside.

Mexicans helped to build railroads such as the Southern Pacific and Santa Fe. The labor conditions and housing were often abominable, as one worker recalled.

It [the house] had one room and a very small kitchen. The ten of us lived there for eight months, in one room. In Mexico we had four rooms. It was a dirty house, and although the Company was to furnish the house there was not a single mattress, only sacks on the beds. All of us could not lie down at once. We slept in turns. We had never done this in Mexico.

A group of Mexican track workers in Arizona in the early 20th century. Men who had helped to build railroads in northern Mexico often emigrated to find similar jobs in the United States.

The grandmother of a migrant family stoops to pick tomatoes.

Farm workers in Rio Grande, Colorado, bring in the harvest of potatoes.

FOOD FOR THE TABLE

The exploitation of Mexican farm workers has been a mainstay of U.S. agriculture in many regions of the country. Few U.S. families, enjoying a meal in the best-fed nation in the world, have ever realized the hardships endured by those who planted and tended the crops. Isabella Mendoza, who was born in Mexico and moved to the United States when she was two, described a lifetime of such work, beginning around 1924.

When I was about eleven or twelve, we all came to California and started to work in the cannery. I worked in the cannery, too, and in the fields picking apples and prunes and that sort of thing; made a little money.

I never learned how to read or write, just to speak English. I went to school a little. Not a real school, but a big room with a table and chairs...but the teacher, she didn't know no Spanish, so she couldn't teach me so good...so I only went to school three or four months, and I can't read nothing to this day.

Then I met up with my husband and we started having kids. I was fifteen when I had the first one. And we worked in the fields—cut apricots, peaches, pears; went to Fresno—picking grapes. My children worked in the fields, too, picking apples. They started before they went to school. My two boys were four and five and they helped us fill the baskets. I used to work till it was time to leave to make something to eat, and then I'd come home and cook for my family, for my husband and the boys. And then I have the baby, a little girl.

You know what I'd do when she was small? I had a tomato box, and my older boy and my other boy, they make a little wagon with the wheel and I'd put the baby and the pillow in that tomato box. And I put her in there and I go out and pick apples from the ground and I pulled her along in the wagon.... The only thing is, when we were working and she was small and she start hollering, "Mamma! Mamma!" "What do you want?" I'd say. "Mamma, I want some chichi." That means she wanted to feed from the breast, you know, milk. And my husband would say, "Go on, feed that kid over there." So I'd leave the can or the basket...and go feed her, give her some milk, and then come back and help him fill the basket.

It was kind of hard because we was always going from one place to another, back and forth, back and forth—apple orchards, lettuce fields, the grapes. Sometimes you'd sleep in the car—you'd be worried about getting killed on the road. Or you'd camp in the dark, and nobody wants you on their land anyhow. We'd be in Riverside and the orange season would end over there, and we'd go to Fresno for the grape season.

We'd travel day and night to get to Fresno. And then tomatoes up at Tracy, so there you'd go. And then plums over here toward Santa Clara; that's where there's a lot of plum orchards. I know this whole country like a book; all the vegetable, all the fruit....

The thing is, traveling around like that, you don't get ahead. The money looks good, but the money would creep away: gas for the car, or you'd break down, you have to get parts for it, see. Then you got to eat, so you'd go to the restaurant with a whole bunch of kids—and all of them were small— it cost money for food in a restaurant and milk and everything. So you don't make nothing much. So, finally, we settle here near Watsonville. There's a cannery here. My husband work with the cannery, and the children would help with the picking in the brussels sprouts and the other fields around here.

Children of the farm workers often did not receive much formal education. A 90-year-old woman in south Texas told a writer how she missed going to school.

Somebody came around to note us down to go to school, but we never did. They never came back to see if we were going to school. I guess nobody was interested in our education, not our parents, and not the school people for sure. Now we are ignorant. Nobody cared then, but we care now. We suffer much for these things. Somebody did us wrong. The school people just wanted that money from the state. We did not know what they were doing then. They took our names and got the state money. We were stupid to let them. But our parents needed us in the fields, so we never went to school. We never learned to read and write.

Law enforcement was always on the side of the employers. In 1921, Elias Garza told sociologist Manuel Gamio his experience.

In San Antonio, we were under contract to go and pick some cotton in a camp in the Valley of the Rio Grande. A group of countrymen and my wife and I went to pick. When I arrived at the camp the planter gave us an old hovel which had been used as a chicken house before, to live in, out in the open. I didn't want to live there and told him that if he didn't give us a little house which was a little better we would go. He told us to go, and my wife and I and my children were leaving when the sheriff fell upon us. He took me to jail and here the planter told them that I wanted to leave without paying him for my passage. He charged me twice the cost of transportation, and though I tried first not to pay him, and then to pay him what it cost, I couldn't do anything. The authorities would only pay attention to him, and as they were in league with him they told me that if I didn't pay they would take my wife and my children to work. Then I paid them.

An Anglo commented on the work in the sugar beet fields:

I do not want to see the condition arise again when white men who are reared and educated in our schools have got to bend their backs and skin their fingers to pull those little beets.... You can let us have the only class of labor that will do the work, or close the beet factories, because our people will not do it, and I will say frankly I do not want them to do it....

If you are going to make the young men of America do this back-breaking work, shoveling manure to fertilize the ground, and shoveling beets, you are going to drive them away from agriculture.... You have got to give us a class of labor that will do this back-breaking work, and we have the brains and ability to supervise and handle the business part of it.

A woman picks spinach on a farm near Robstown, Texas.

Mexican American farm workers box apricots near Canoga Park, California, in the 1920s.

Particularly in Texas, Mexican laborers were hired to pick cotton, brutally hard and backbreaking work. The contratista *(contractor), usually Mexican himself, recruited workers for the task. A daughter of a* contratista *described the conditions in the cotton fields of southern Texas in the 1920s.*

We would go to the field very early in the morning, like five o'clock. That was the ideal time to pick cotton because it was still wet from the dew and thus heavier. [The workers turned their bags in at the end of the day, and were paid by how much the cotton weighed.] We wanted to start working right away, but the owner would not let us. My father usually brought a truck load of people, families from the neighborhood that would come to work with us. He was responsible for them and would receive a proportion of what they procured. When the patron told us to start working, we would fall on those fields like busy bees. I worked as fast as I could. Sometimes my father would come over and help me in my work by filling my bag. I enjoyed that. What I didn't like was when he helped other people who did not belong to our family. I would ask him, "Why do you help them?" And he'd say, "Because it is good to help others; if I am good to them they will be good to me." And it was true, those people were very loyal to my father and always came back to work with him and for no other *contratista*. About noon, we'd all stop to eat our rolled tortillas. It would take us no longer than fifteen minutes to eat lunch. My father brought fresh water in great wooden barrels, and we could drink from it once in a while. After noon, the work became slower and heavier, because the cotton was dry, and it would take us longer to make a 100-pound bag, for which we were paid.

A child farm worker in Texas waits for the overseer to tally the sugar beets he has picked.

Labor contractors exercised great power over the Mexican workers, as Ernesto Galarza explained.

There was never any doubt about the contractor and his power over us. He could fire a man and his family on the spot and make them wait days for their wages. A man could be forced to quit by assigning him regularly to the thinnest picking in the field. The worst thing one could do was to ask for fresh water on the job, regardless of the heat of the day; instead of iced water, given freely, the crews were expected to buy sodas at twice the price in town, sold by the contractor himself. He usually had a pistol—to protect the payroll, so it was said. Through the ranchers for whom he worked, we were certain that he had connections with the *Autoridades,* for they never showed up in camp to settle wage disputes or listen to our complaints or to go for a doctor when one was needed. Lord of a rag-tag labor camp of Mexicans, the contractor, a Mexican himself, knew that few men would let their anger blow, even when he stung them with curses.

Raul Morin recalled how the depression of the 1930s forced many Mexican Americans back into the fields.

The average United States Mexican laborer was classified in the unskilled or semi-skilled category. Many who held good paying jobs in cities and towns had no other choice but to turn back to farm work and stoop labor in the field—to work in the much-hated jobs that for years, since the early border-crossing days, had been classified as *"suitable for cheap Mexican labor only."*

Our educational progress was retarded everywhere when youngsters of high school age were forced to drop out of school to seek work and help the family earn a living. Many of us joined the CCC [Civilian Conservation Corps] forestry camps, others worked in NYA [National Youth Administration] and many took to the road roaming up and down the country, seeking work or just leaving to lessen the burden at home."

The bracero program (1942–1964) brought Mexican laborers to the United States, primarily to work in agriculture. Although the United States guaranteed the government of Mexico that the braceros would receive decent wages and working conditions, the laborers were usually at the mercy of those who hired them. One bracero said:

The most galling part of the whole process is the uncertainty of it all. Men have to wait on the scene for weeks. Day after day there is no hiring. Some days hundreds of men must stand for hours under the broiling sun while the representative of some California grower selects some fifty from their number. Then there is another indefinite wait until another rancher arrives.

In the 1990s George Garza of Florida City, Florida, described labor contractors:

They're thieves. They get a dollar five a bucket for tomatoes, and they pay the farmworkers forty cents. The contractors say it's for bookkeeping, taxes, social security, workmen's comp., health insurance, but they don't pay anything; they're just thieves.

A Mexican contratista, *or labor contractor, in Texas in 1939.*

As Others Saw Them

Growers in states as far north as Michigan profited from the work of Mexican farm laborers. The Michigan Department of Social Welfare reported on the housing conditions of migrant workers in the 1940s:

These people are living in a colony of dilapidated shacks on the property of the sugar company and inside the corporation limits of the village. The majority of them consist of only two small rooms and are in a filthy condition full of vermin and without sanitation.... Each of these houses is occupied by one to three families. In one instance there were fourteen people and only one bed. There is an inadequate number of outdoor toilets. None of them has probably ever been cleaned or limed.

A boy and his mother pick carrots on a Texas farm. Because their work helped support the family, few children of farm workers ever attended school—dooming them to continue their parents' lives as unskilled, poorly paid workers.

José Garcia, who first arrived from Mexico in 1959, was part of the bracero program.

At that time, you know, you could earn from seven to ten pesos [a peso was worth about eight cents] a day in Mexico, whereas here you could earn from sixty to seventy dollars a week. And we were poor. Our whole family was poor. It was the money, that's why I came.

At that time they had a law about the *braceros*. You were allowed to cross the border with a permit, a work permit, for a certain length of time: sometimes sixty days, sometimes ninety days. As soon as we'd cross the border, we'd go to a place in Texas like a big hall, a dance hall, and the bosses would pick out maybe ten or twenty to go to this town, ten or twenty to go to another town. And then they would ship us out....

The food was pretty poor. Really just rotten. They cooked everything mixed together. Carrots, turnips, corn, peas. And in order to find a piece of meat you needed a magnifying glass. The barracks were dirty, completely dirty. It was just awful. I feel kind of funny telling you, but the bathrooms—what you call them?—the outhouses were just awful. The board would be like maybe seven feet long, and in those seven feet they would have three holes; and the board would be covered, completely covered, with worms, white worms—those great giant white ones. That's why, to this day, I won't look at white rice. I won't eat no white rice because it looks exactly like those worms, you know. At that time a lot of us would not eat the white rice that they served us, because we said the cook used to go out to the toilet to get it.

In 1977, Elizabeth Loza Newby recalled growing up in a migrant worker family.

Until I was 14 years old, I was never conscious of sleeping anywhere except in the back of a truck. My father...decided to...join the migrant circuit in the spring of 1948, when I was 16 months old. He invested all of his money in a 1942 army surplus truck, which was to become our home for the next 13 years. He cleaned and repaired the bed of the truck to make it suitable for living. Orange crates were used for stands and cupboards and baby beds (my first bed was an orange crate). An old-fashioned metal tub filled with coal was our stove....

As I outgrew my orange crate, my bed was transferred to the floor, where three or four hand-sewn quilts from Mexico provided me with a mattress. Since there were no partitions separating my parents from my brothers and me, we had to learn to get along together....

Compared with the migrant housing provided by many of the farmers for whom we worked, ours was luxurious. As I grew older, I helped my mother keep the truck spotlessly clean; and, by camping away from the rest of the migrants, we escaped the epidemics of tuberculosis, influenza, measles, and other contagious diseases....

For my family, each new day brought both new adventure and new problems. There was the time in Nebraska when, after we had finished hoeing several acres of sugar beets, the owner informed us that he wasn't pleased with our work. Rather than have us do it over, he simply refused to pay wages.... Of course we needed the money to buy food and supplies to keep moving, so this setback caused real hardships. Since my parents, along with other migrants, were...ignorant of their rights...we just moved on to the next job and prayed that this type of incident would not be repeated....

There was the time in Oklahoma when we were picking cotton. All of a sudden a crop-dusting plane descended upon the field in which we were working, spraying it with insecticides. Dad heard the plane coming and...yelled to everyone to go to their vehicles and close the windows. Though most of them did, a few did not; and they soon became ill from exposure to the insecticide.

In 1992, 60-year-old Rita Pérez of Idaho Falls recalled working in the fields in 1940, when she was only eight years old. Before her birth, her parents had moved from Mexico to Idaho.

My father...started the year thinning beets. Then we would go to Driggs and pick green peas and come back [to Idaho Falls] about mid-September. After that, the potato harvest. If there were any sugar beets to harvest, we would also work at that....

My parents would take us in the car and park it at the end of the field. My mother could come at intervals to check on the baby and nurse him. The first time I was left in charge, my younger sister was about six weeks old and I was not quite five.

The first time I worked out in the field, I was about eight years old. We worked from sunrise to sunset...picking green peas.... I believe we got paid three cents per pound.

Although Policarpo Castro had been a skilled stonemason in Mexico, he was forced to take unskilled farm jobs after immigrating to the United States in the 1920s. Castro commented:

I have gone from one place to another working as a laborer for I haven't found anything else because the masons' union don't want to admit Mexicans.... But although I have worked as a laborer I have always tried to learn everything that I could. I have worked in cement, in a brick-yard, laying pipes...and have learned all that sort of work, even how to make entrances and walks for a garage with an incline. All that will do me some good in Mexico.... I know that if I want to amount to something in any work I will have to do it there in Mexico, because the Americans only despise us.

Mexican farm workers pick cotton on a Texas plantation in 1919.

A label on a box of avocados ridicules the Mexican American laborers who did the hard work of picking the fruit.

PREJUDICE

Lupe Nieto came to the United States with her family during the Mexican Revolution. They settled first in Santa Ana, California, then moved to the San Joaquin Valley. Years later, Nieto still flushed with anger when she remembered the prejudice she faced.

I was born in Mexico, but I was only a year and a half when we came to this country. I didn't become an American citizen until 1958. We were raised here in California.

For a long time I was very bitter; I felt bitter. A neighbor wouldn't let her kids play with my kids because we were Mexicans, and she claimed that my children had lice in their hair. Why, I may be Mexican but I'm not dirty! And I didn't like that woman since then. That was the only woman in Firebaugh that ever made me feel that way.

[They were looked down on in many ways. When they tried to rent an apartment, they were told,] "Sorry, we don't rent to Mexicans." And in some theaters they would not allow Mexicans. I'm the darkest in my family. I have a brother and sister who have blue eyes and light brown hair. You could swear they aren't Mexican. Yet they weren't allowed in a theater. A couple of times we were not served in a restaurant. When somebody does that to me—oh, I can be a demon! And the only reason I didn't say anything was because I didn't want to embarrass [my husband]. I would have embarrassed him to tears. But I wanted to lash at them! We sat there and sat there. People came and they served them, and they *never* served us and they never asked us! So, what more of a hint do you want?...

That grew up with me for a long, long time. It wasn't until just about ten years ago that it left me. But it does leave you with that ugly feeling that you're inferior.

The Ku Klux Klan, which terrorized African Americans in the southern United States, also targeted Mexican Americans. This Klan meeting took place in Santa Barbara, California, around 1921. The local Klan tried to drive Mexicans out of the area because they were Catholics, another group that the Klan opposed.

Sotero H. Soria arrived in the United States in 1913, staying with his brother in Garden City, Kansas. Years later, as Soria looked back on his life, he recalled the prejudice against Mexican Americans.

There were some six or seven [Mexican] families here when I came. It was not easy for us; we suffered a lot. In spite of everything, from time to time, we got together. Although we were poor, we danced and played our guitars so that we could enjoy a little of life.

We, the old ones bought or rented houses in this district on this side of the railroad track because they would not sell or rent us houses on the other side. There were men who didn't want us here.

When my son went off to war [World War II], I went with him to a bar. I asked for a beer. The owner told me that he would sell it to me but that I would have to drink it outside because they wouldn't allow me to drink it inside.

"Listen, my son is leaving for Germany to fight for us so that you can have your business."

"Get out, get out!" was the proprietor's answer.

It has been different for the new generation. We don't want them to suffer. We want them to have a better life in the United States.

José Garcia, on his second trip to the United States, had immigration papers that allowed him to get a job in a mushroom plant. He described how Mexicans were treated in the 1970s.

Of course, they favor Anglos in the plant. They give them the easier jobs. They don't like the Mexicans to touch the machinery, you know. Where I work, if they had the custom of carrying things on a burro, they wouldn't let the Mexicans touch the burro. They would only let the Anglos touch the burro. They don't like us to touch things. They think we're all from the hills. The Anglos run the rollers to carry the straw and the compost, and just move the levers back and forth to make the machines work. But the Mexicans, the majority of the Mexicans pick mushrooms, plant them by hand—it's the Mexicans. That's the kind of work they think they're good for....

They hurt you in words, too. The son of the main boss has a pickup, a green pickup, and it's fairly new. And sometimes the workers go over there and they lean against it when they're on their lunch or their break, and that kid will come up and say, "Hey, what's going on here? You guys having a meeting or something? Why are you leaning against my pickup?" He thinks his car is too good for us to lean against.

Still, things are getting better here. When I first came [in 1959], there were some stores and restaurants that wouldn't even serve a Mexican. They'd say, "We're just closing," or "We don't have any more food," or sometimes just, "Get out." They don't do that anymore.

I want to save money and buy a ranch here.... That would

The Sleepy Lagoon Case

Defendants in the Sleepy Lagoon trial.

Early on the morning of August 2, 1942, a young man named José Diaz was found lying on a dirt road on the outskirts of Los Angeles. Diaz was taken to a hospital but died of a skull fracture without regaining consciousness. At the time, the Los Angeles newspapers had been playing up the Mexican "crime" problem with stories about fights between groups of gangs. They welcomed the case, dubbing it the "Sleepy Lagoon" case after the reservoir near the spot where the body was found. The police rounded up hundreds of Mexican American youths. Twenty-two were tried for conspiracy to commit murder.

From the beginning, the criminal proceeding was unfair. The grand jury heard a document written by Ed Duran Ayres of the Los Angeles sheriff's department that stated, in part, that "this Mexican element feels a desire to kill or at least draw blood." Ayres attributed this to an "inborn characteristic that has come down through the ages."

At the trial, moreover, the defendants were not allowed to sit with their lawyers and could only speak with them during recesses in the proceedings. At the order of the prosecutor, they were not allowed haircuts and fresh clothes, so they appeared before the jury looking like hobos. During the three-month trial, not one witness gave evidence that placed any of the defendants near José Diaz at any time. Even so, 17 of them were convicted on charges from first-degree murder to assault with a deadly weapon.

Immediately, a Sleepy Lagoon Defense Committee arose to get new lawyers and to appeal the case. Two years later, the convictions were overturned. The appeals court reprimanded the trial judge for his unjudicial behavior and the defendants were freed after spending more than two years in jail. The president of Mexico praised the overturning of the verdict, but the trial and the publicity surrounding it showed the kind of prejudice that Mexican Americans had to endure.

be my dream. Yeah, my ambition, my dream, if I could have a ranch. I'd like raising animals there—chickens, pigs, cows, whatever is easier to sell, whatever there is a demand in. That's what I'd like to do if I could only get the money together.

The forces of law and order were seldom on the side of Mexican Americans. In the 1980s, Frank Escalante recalled the lack of justice for Mexican Americans in the Tucson Basin of Arizona.

I don't want you to think that I am prejudiced, but the facts are the facts. The rich have always stomped on the poor. I know some people who had more than 300 acres of land this side of the mountain. They worked for a rancher who told them that because they worked for him, he would pay the taxes on their land. And when he paid their taxes, he told them to get off the land. Just like that....

My dad used to tell me stories about how some of the people were said to have lost their land. For instance, the Arizona Rangers. You remember how famous they used to be? Well, it's said they had a little trick. Say first someone wanted a little ranch. He'd go up to the ranch and put hides in the corral and then accuse the man of rustling. My mom's dad used to live on the Leon's Ranch above Loma Alta where it happened. He knew a man it happened to. But the man was wise to what the Rangers were up to, and when he found he was surrounded he crawled away from the ranch. He went to my grandfather's to borrow a horse. They shod the horse by lantern light. He wrote to my grandfather later that by sunrise he was in Santa Cruz, Mexico. He had to forfeit his land, but if he had stayed he would have been hanged.

César Chávez, who later organized the migrant farm workers into a union, recalled the discrimination he experienced while growing up in the town of Brawley, California.

We used to shine shoes, my brother and I, and do any thing to make a dime, and we saved tin foil from cigarette packs to trade for shoes; we really hustled. The cops wouldn't let us into Anglo Town where the white people lived, but there was a diner right on the line, kind of, and everybody talked about how it was supposed to have beautiful hamburgers.

It also had a sign reading "Whites Only," but we had just come from the country, from Arizona, from a community that was mostly Mexicans or whites too poor to bother about us. So we didn't understand and went in.

The counter girl was up at the far end with her boyfriend and I said "Two hamburgers, please."

The girl said, "What's the matter, can't you read? Damn dumb Mex." She and her boyfriend laughed, and we ran out. Richard was cursing them, but I was the one who had spoken to them and I was crying. That laugh rang in my ears for twenty years. It seemed to cut us out of the human race.

School could be an unpleasant experience for Mexicans. Martha Morales described her experience.

We came in June 1923.... And we tried to learn English as fast as we could, which we did.... It was no disgrace to be poor.... We were Mexicans, there was no way to beat it. In fact our skin color is different, our cultures were different. [We had] to accept it and be proud.... That's the way we were raised.

So when we went to school if anybody called us greasers, or [said], "You eat tortillas"...they were saying the truth. Because we learned not to get mad. It hurt...but we just grew up that way.

I do remember...my teacher did tell me not to wear my bow to school. And it kind of broke my heart because I had been raised with a bow in my hair.... We tried to excel in our studies. That was our way of recompensing.... Even though we were proud to be Mexican, we figured they'd leave us alone about tortillas and frijoles if maybe we can do better than the next person.

Raul Morin, today a Mexican American scholar, recalled the prejudice he faced as a boy.

By virtue of having been born in the United States, we began on equal footing with other Americans. After a normal happy childhood, we ran into our first little surprise. We learned, first from our parents, then from our scrupulous Anglo neighbors, that we were of a different breed...we were *Mexicans*. We calmly accepted this fact, then went along with the teachers...in school to learn Americanism: the English language, American History, laws and customs of this country. Surprise and puzzlement came at home when we tried to put into practice the things we had learned at school. We were often reprimanded, scolded and laughed at for such things as speaking the English language, refusing to take the food we ate at home for school lunches, and for changing our name from José to Joe.

Prejudice against Mexican Americans in Texas was particularly bad. An elderly woman remembered:

Sometimes *mi querido* (my husband) would come home tired and angry. I would talk about the *bolillos* (Anglos) and how bad they were. He wondered if maybe we Mexicans were as bad as they said. I think my husband didn't think he was much good. He thought the *bolillos* were better than us, but he never said that. He couldn't say such a thing because he hated them too much. If he said that, he would not work as good. There was just too much...I don't know...too much work I guess. It wasn't something we talked about much, you know. I guess we didn't know how to or something. But I loved my husband. He was a good man.

The Zoot Suit Riots

Mexican Americans were rarely protected by the police and indeed were often the victims of police brutality—a problem that remains today. Never was this more clear than in the so-called zoot suit riots, named for the clothing favored by young Mexican American males in the early 1940s.

On June 3, 1943, when many Mexican Americans were serving in the military during World War II, some sailors ventured into a Mexican neighborhood in Los Angeles and got into a fight. No one ever found out who had attacked them. But the next night, carloads of sailors prowled the barrios of Los Angeles looking for Mexicans to beat up. The police did nothing about these attacks, but the city's newspapers whipped up a frenzy of anti-Mexican fever, blaming Mexican Americans for the violence. The attacks continued for a full week before the Military Police and Shore Patrol declared downtown Los Angeles out of bounds for military personnel.

An honest Anglo journalist, Al Waxman, described what he witnessed on June 7:

At Twelfth and Central I came upon a scene that will long live in my memory. Police were swinging clubs and servicemen were fighting with civilians....

Four boys came out of a pool hall. They were wearing the zoot-suits that have become the symbol of a fighting flag. Police ordered them into arrest cars. One refused. He asked: "Why am I being arrested?" The police officer answered with three swift blows of the night-stick across the boy's head and he went down. As he sprawled, he was kicked in the face. Police had difficulty loading his body into the vehicle because he was one-legged and wore a wooden limb. Maybe the officer didn't know he was attacking a cripple.

At the next corner a Mexican mother cried out, "Don't take my boy, he did nothing. He's only fiteen years old. Don't take him." She was struck across the jaw with a night-stick and almost dropped the two and a half year old baby that was clinging in her arms....

I came upon a band of servicemen making a systematic tour of East First Street.... Three autos loaded with Los Angeles policemen were on the scene but the soldiers were not molested. Farther down the street the men stopped a streetcar, forcing the motorman to open the door and proceeded to inspect the clothing of the male passengers. "We're looking for zoot-suits to burn," they shouted. Again the police did not interfere.

Jacinto and Carmen Orozco with their family in Tucson, Arizona, around 1940.

CHAPTER FIVE

PUTTING DOWN ROOTS

The majority of Mexicans who settled in the United States after the 1880s clustered in *barrios*. A barrio might be as small as a cluster of shacks along a railroad track or as large as East Los Angeles, a sprawling urban area with a population of more than 100,000. Often named after the area of Mexico where their residents came from (like Chihuahita, meaning Little Chihuahua), the barrios provided a sense of security. There, Spanish was the predominant language; stores sold *la comida*, familiar foods; and the customs of the old country were preserved.

Local governments frequently treated the barrios with the same neglect accorded to neighborhoods of other people of color, such as African Americans and Asian Americans. Proper water systems, sewage facilities, and electrical utilities came last to the barrios, when they came at all.

Many Mexicans, driven north by poverty or by the violence of the Mexican Revolution, regarded their stay in the United States as a temporary one. Even if parents had to learn English for their jobs, they spoke Spanish at home to their children. They continued to celebrate the Mexican patriotic holidays, such as September 16 (Mexico's Independence Day) and Cinco de Mayo.

In the southwestern United States, where the Roman Catholic clergy was dominated by French and Irish priests, Mexicans at first found little respect for their traditional forms of religion.

Nonetheless, the majority of Mexican Americans remained faithful to Catholicism, remembering that Hidalgo, the leader of the Mexican independence movement, had been a priest. The faith was part of the daily life of Mexican American communities. Even when priests and churches were far away, families nourished the spirit of religion in the home. Religious expressions, such as *Gracias a Dios* (Thanks be to God) and short prayers to the Virgin of Guadalupe, were everyday customs.

People went to church to commemorate important moments of their lives: First Communion, confirmation, marriage, and funerals. The church calendar punctuated the year from the penitential season of Lent to the joyous celebration of Christmas. Mexican Americans celebrated the feast day of their patron saint (after whom they were named) in addition to their birthdays.

However, relatively few Mexican Americans became priests, in part because they received little encouragement from the church hierarchy. Despite the huge number of Mexican American Catholics, none was consecrated as a bishop until 1970, when Patrick Flores, the son of migrant workers, became auxiliary bishop of San Antonio, Texas. Nine years later, he became the first Mexican American archbishop.

The few Mexican American immigrants who rose to middle-class status before 1945 usually did so by founding stores and other businesses that catered to the barrio. Even these economically successful people often found themselves rejected by the Anglo community. Many, especially in Texas, were forced to send their children to segregated schools, where the teachers mentioned Mexicans only as a conquered people.

Yet the Mexican Americans' strengths helped them survive. The close-knit family, part of both the Spanish and the Native American heritage, provides warmth, emotional support, and help in times of need for its members. *La Familia* is an extended one that includes grandparents, aunts, uncles, and cousins.

Traditionally, the father of the family is a strong figure who assumes responsibility for "all under his roof." Children are expected to show *respeto* (respect) for elders. They usually have a very affectionate relationship with their grandparents, *el abuelo* and *la abuelita*.

When a child is born, its parents establish an important relationship called *compadrazgo*, by choosing close friends or family members as sponsors for the child's baptism. The godparents, or *padrino* and *madrino*, are responsible for the child's upbringing if the parents die. This relationship between child and godparent lasts for life.

Equally strong is the relationship between the godparents and the parents of the child. They call each other *compadres* or *comadres*, literally "co-parents," and are expected to help each other in time of crisis or need. In the United States, the *compadrazgo* relationship helped migrants obtain jobs and places to live and gave them help in getting a start. Often, a young father would enhance his status in the community by choosing an older and more respected man to become his compadre.

Mexican cultural tradition drew a sharp line between the conduct of the sexes. Young women were carefully chaperoned, for the promiscuity of a daughter or sister brought disgrace on the whole family. Young men, on the other hand, grew up respecting the ideal of *machismo*, which comes from the Spanish word *macho*, or "male."

In the urban barrios of the United States, young men formed gangs as a way of asserting their self-worth and dignity. In Los Angeles, Mexican gangs began to appear as early as the 1920s. Of course, this phenomenon was not limited to Mexican immigrants; Irish and Italians, among others, also formed neighborhood gangs in the United States. Moreover, the Mexican American gangs were at

Pantaleón Duran in the doorway of his grocery store in a Los Angeles barrio in the 1920s.

first relatively benign organizations. In the 1940s a committee headed by a Catholic bishop of Los Angeles reported that "many of [the gangs] are wholesome social groups, meeting in public schools, in sub-police stations, and under the supervision of responsible officials."

More formal kinds of organizations arose in areas where Mexican Americans had lived since before the U.S.–Mexican war. In the late 19th and early 20th centuries, middle-class Mexicans formed *mutualistas*, groups that provided benefits such as life insurance, or-

ganized social and cultural events, and also fought for Mexican American rights. The most important early *mutualista* was the Alianza Hispano Americana, founded in Tucson, Arizona, in 1894. Its members included middle-class lawyers and ranchers who were descendants of the "original" Mexican Americans. Lodges of the Alianza soon multiplied throughout the Southwest. By 1929 the Alianza had 15,000 members as far north as Chicago and as far west as San Diego.

The word *Hispano* in the title of this organization was significant, for the older Mexican Americans did not feel a strong kinship to the newer immigrants. The "original" Mexican Americans stressed the Spanish elements of their heritage, often calling themselves Hispanos, rather than Mexicans. However, other *mutualistas*, like La Sociedad Fraternal Moctezuma and La Sociedad Benito Juárez, were founded by the recent immigrants themselves and stressed their Mexican heritage.

A new, more politically active type of group, the Liga Protectora Latina, was founded in Phoenix in 1914. Its members, local business leaders of the Mexican American community, were angered by a bill introduced in the Democrat-controlled Arizona state legislature to bar non-English-speaking people from "hazardous occupations." The proposed law was seen as a way to prevent Mexicans from working in the state's mines. The

Liga successfully lobbied against passage of the law, and its growing political power helped elect an Anglo Republican to the governorship of Arizona in 1918.

The previous year, the United States had entered World War I against Germany. One reason for U.S. involvement in the war was the Zimmermann telegram, in which Germany's foreign minister proposed an alliance with Mexico. The German minister offered "an understanding on our part that Mexico is to reconquer her lost territory in Texas, New Mexico, and Arizona."

Though Mexico did not accept the offer, Mexican Americans in this country feared that they would be suspected of disloyalty. Groups such as the Alianza and the Liga staged patriotic parades and encouraged Mexican Americans to cooperate with the military draft. (Though relatively few immigrants were U.S. citizens, they were required to register at the draft board offices anyway.) The support of the *mutualistas* for the war effort heightened their political influence, and after the war they continued to grow in size and importance.

The arrival of large numbers of Mexican immigrants during the 1920s resulted in an enormous growth in the barrio population. Mexican American leaders began to protest the racism that the newcomers faced. In 1929, the League of United Latin American Citizens (LULAC) was founded in Corpus Christi, Texas. Its leaders, primarily younger Mexican Americans who had been born in the United States or arrived as children, declared their intention "to use all the legal means at our command" to ensure "equal rights, the equal protection of the laws of the land and equal opportunities and privileges" for all citizens.

Dolores Curiel Tapia Zamora, who lived on a Texas ranch, poses in front of an American flag during World War I. Because Germany tried to make an alliance with Mexico during the war, Mexican Americans publicly demonstrated their loyalty to the United States.

LULAC provided legal help for Mexican Americans accused of crimes; filed lawsuits to overturn segregated schools and other public facilities; and fought to gain the right to vote and to serve on juries wherever Mexican Americans were denied these rights. Today, it is a national organization, with more than 110,000 members, that has continued to be in the forefront of Mexican Americans' efforts to obtain their rights.

The founding of LULAC came just before the beginning of the Great Depression of the 1930s, a time when all Americans suffered from economic hardship. But the depression hit Mexican American communities particularly hard, for it led to the repatriation program that shattered life in the barrio by sending whole families back to Mexico.

However, some of the families that had arrived since 1910 had made the decision to become citizens of the United States. Their children graduated from high school in growing numbers, and many of them benefited from the New Deal programs that were designed to employ young people and keep them above the poverty line. Mexican Americans took another step forward in 1935, when Dennis Chávez, a New Mexico Democrat, became the first U.S. senator of Mexican descent.

When the United States entered World War II in 1941, young Mexican Americans demonstrated their patriotism by joining the armed forces. In proportion to their numbers in the general population, more Mexican Americans served than any other ethnic group. Seventeen of them would win the Congressional Medal of Honor, the nation's highest award for valor.

The national war effort created opportunities for Mexican Americans, both men and women, who found jobs in many industries where they had never been allowed to work before. After World War II, these workers, along with returning veterans, began to assert the rights that had long been denied them.

EL BARRIO

Most Mexican immigrants settled in their own neighborhoods and communities, called barrios. *Mario Suarez, a onetime migrant laborer who became a writer, described a barrio in Tucson, Arizona, called El Hoyo.*

Its inhabitants are chicanos who raise hell on Saturday night and listen to Padre Estanislao on Sunday morning. While the term chicano is the short way of saying Mexicano, it is not restricted to the paisanos who came from old Mexico with the territory or the last famine to work for the railroad, labor, sing, and go on relief. Chicano is the easy way of referring to everybody.... However, it is doubtful that all these spiritual sons of Mexico live in El Hoyo because of its scenic beauty—it is everything but beautiful. Its houses are simple affairs of unplastered adobe, wood, and abandoned car parts. Its narrow streets are mostly clearings which have, in time, acquired names. Except for some tall trees which nobody has ever cared to identify, nurse, or destroy, the main things known to grow in the general area are weeds, garbage piles, dark-eyed chavalos [kids], and dogs. And it is doubtful that the chicanos live in El Hoyo because it is safe—many times the Santa Cruz [River] has risen and inundated the area.

In other respects living in El Hoyo has its advantages. If one is born with a weakness for acquiring bills, El Hoyo is where the collectors are less likely to find you. If one has acquired the habit of listening to Octavio Perea's Mexican Hour in the wee hours of the morning with the radio on at full blast, El Hoyo is where you are less likely to be reported to the authorities. Besides, Perea is very popular and sooner or later to everybody "Smoke in The Eyes" is dedicated between the pinto beans and white flour commercials. If one, for any reason whatever, comes on an extended period of hard times, where, if not in El Hoyo are the neighbors more willing to offer solace? When Teofila Malarara's house burned to the ground with all her belongings and two children, a benevolent gentleman carried through the gesture that made tolerable her burden. He made a list of five hundred names and solicited from each a dollar.... When the new manager of a local store decided that no more chicanas were to work behind the counters, it was the chicanos of El Hoyo who, on taking their individually small but collectively great buying power elsewhere, drove the manager out and the girls returned to their jobs.... When someone gets married, celebrating is not restricted to the immediate friends of the couple. Everybody is invited. Anything calls for a celebration and a celebration calls

El barrio *could be a separate section of a large city. This is Olvera Street in Los Angeles in the 1940s. Though the city hall tower can be seen in the distance, life in the barrio was in some ways similar to life in a Mexican village.*

Two Mexican American boys at a fiesta in Penasco, New Mexico, in 1940.

for anything. On Armistice Day there are no less than half a dozen good fights at the Riverside Dance Hall. On Mexican Independence Day more than one flag is sworn allegiance to....

Perhaps El Hoyo, its inhabitants, and its essence can best be explained by telling a bit about a dish called capirotada. Its origin is uncertain. But, according to the time and the circumstance, it is made of old, new or hard bread. It is softened with water and then cooked with peanuts, raisins, onions, cheese, and panocha [corn]. It is fried with sherry wine. Then it is served hot, cold, or just "on the weather" as they say in El Hoyo. The Sermenos like it one way, the Garcias another, and the Ortegas still another. While it might differ greatly from one home to another, nevertheless it is still capirotada. And so it is with El Hoyo's chicanos. While being divided from within and from without, like the capirotada, they remain chicanos.

Juanita Zazueta Huerta was born in Firth, Idaho, in 1918. Her parents had come to the United States as contract laborers during World War I. In 1992, she recalled her childhood.

My mother used to tell me that some of the contracted Mexican families went back after the war. Others did not. After my Dad got a job on the railroad, we figured that was a steady job, so we just stayed in Firth....

When my younger sister was born, the ladies at the relief society came to our house. They checked on my mother and asked if there was something they could do for her.... They were very kind to us.

Every Christmas these same women would always bring us presents. As children, we were puzzled because they would say,

The Mexican folk tradition includes a belief in mal de ojo, *the "evil eye." Someone with a powerful personality may—even without wanting to—cause illness in others who are weak. A Mexican American in south Texas described such an incident:*

Last week my cousin's cute little baby had *ojo.* In the hours of the late afternoon my cousin was holding her child out in their front yard. One of the neighborhood men returning home from work stopped to talk with the couple, remarking on the child's cuteness. Then he went on his way. That night when they put the child to bed, it began to cry and remained inconsolable through most of the night. Even though I slept in the other part of the house I could hear them moving about with the child. Very early the next morning I could hear my cousin leave the house as she went next door to speak to our neighbor who is a relative of the man with whom they had chatted the evening before. She asked the neighbor to do her the favor of requesting the man to stop at our house on his way to work. The neighbor went and roused the man with strong eyes. On his way to work that man stopped at our place and went into the other room where he ran a hand over the child's face and forehead, cooing to her and talking to her. He didn't remain any longer than about five minutes. When he left the baby had stopped crying.

Aurelio Dominguez (at left) in his grocery store on Solano Avenue in Los Angeles in 1925. Mexican American entrepreneurs like Dominguez catered to the growing population of urban barrios by offering herbs and foods that were unavailable in Anglo markets.

Girls jumping rope in a barrio in Taos, New Mexico.

Two Mexican American women in San Antonio, Texas, in 1939.

"Santa Claus left this at our house." We asked, "How come he never comes here? He could very well have brought those presents here."...

The Mexican American community celebrated *El Diez y Seis de Septiembre* [Mexican Independence Day, September 16].... The celebrations were nice in those days. The people would give speeches on the *Cinco de Mayo* [May 5] and *El Diez y Seis de Septiembre*. Especially the very old proud men and women who were real patriotic with their pictures and pins and red, white, and green ribbons. They always had a man and woman who were leaders. My *comadre* was *la presidenta*. For years they called her *Doña María la presidenta*. Everyone knew her. She gave her speeches in Spanish....

When I was young, I did not really keep up with the speeches. We were itching to start dancing. It was nice; we enjoyed all of it.

Ernesto Galarza described the barrio around Fifth Street in Sacramento, California, between 1910 and 1920.

For the Mexicans the barrio was a colony of refugees. We came to know families from Chihuahua, Sonora, Jalisco, and Durango. Some had come to the United States even before the revolution, living in Texas before migrating to California. Like ourselves, our Mexican neighbors had come this far moving step by step, working and waiting, as if they were feeling their way up a ladder....

[The barrio was] a neighborhood of leftover houses. The cheapest rents were in the back quarters of the rooming houses, the basements, and the rundown rentals in the alleys....

Barrio people...cut out the ends of tin cans to make collars and plates for the pipes and floor moldings where the rats had gnawed holes. Stoops and porches that sagged we propped with bricks and fat stones. To plug the drafts around the windows in winter, we cut strips of corrugated cardboard and wedged them into the frames. With squares of cheesecloth neatly cut and sewed to screen doors holes were covered and rents in the wire mesh mended.

Such repairs, which landlords never paid any attention to, were made *por mientras,* for the time being or temporarily. It would have been a word equally suitable for the house itself, or for the barrio. We lived in run-down places furnished with seconds in a hand-me-down neighborhood all of which were *por mientras....*

The older people of the *barrio,* except in those things which they had to do like the Americans because they had no choice, remained Mexicans. Their language at home was Spanish. They were continuously taking up collections to pay somebody's funeral expenses or to help someone who had had a serious accident. Cards were sent to you to attend a burial where you would throw a handful of dirt on top of the coffin and listen to tearful speeches at the graveside. At every baptism a new *compadre* and a new *comadre* joined the family circle.

New Year greeting cards were exchanged, showing angels and cherubs in bright colors sprinkled with grains of mica so that they glistened like gold dust. At the family parties the huge pot of steaming tamales was still the center of attention, the *atole* [a drink made of cornmeal] served on the side with chunks of brown sugar for sucking and crunching. If the party lasted long enough, someone produced a guitar, the men took over and the singing of *corridos* began.

Oscar Zeta Acosta, in his Autobiography of a Brown Buffalo, *recounted his memories of childhood in a California barrio during World War II.*

Although I was born in El Paso, Texas, I am actually a small town kid. A hick from the sticks, a Mexican boy from the other side of the tracks. I grew up in Riverbank, California; post office box 303; population 3,969.... The sign that welcomes you as you round the curve coming in from Modesto says, "The City of Action."

[My father] Manuel Mercado Acosta is an *indio* from the mountains of Durango. His father operated a mescal distillery before the revolutionaries drove him out. He met my mother while riding a motorcycle in El Paso.

Juana Fierro Acosta is my mother. She could have been a singer in a Juárez cantina but instead decided to be Manuel's wife because he had a slick mustache, a fast bike and promised to take her out of the slums across from the Rio Grande. She had only one demand in return for the two sons and three daughters she would bear him: "No handouts. No relief. I never want to be on welfare."

We lived in a two-room shack without a floor. We had to pump our water and use kerosene if we wanted to read at night. But we never went hungry. My old man always bought the pinto beans and the white flour for the tortillas in 100-pound sacks which my mother used to make dresses, sheets and curtains. We had two acres of land which we planted every year with corn, tomatoes and yellow chiles for the hot sauce. Even before my father woke us, my old ma was busy at work making the tortillas at 5:00 a.m. while he chopped the logs we'd hauled up from the river on the weekends....

Bob [his brother] and I had to chop wood for the evening meal. We had to pump the water into tin tubs for our nightly bath. And unless we bathed and washed the dishes, we couldn't turn on the little brown radio to listen to *The Whistler, The Shadow,* or *The Saturday Night Hit Parade* with Andy Russell, the only Mexican I ever heard on the radio as a kid. We would sit and listen while we shined our shoes. During the commercials my mother would sing beautiful Mexican songs, which I then thought were corny, while she dried the dishes. "When you grow up, you'll like this music too," my ma always prophesied.

Felix Arnelas and Margaret Martinez at a beach in San Pedro, California, in the 1930s. Her bathing suit would have been regarded as too daring in Mexico, but immigrants adjusted to the social freedom found in the United States.

A woman milks a goat in the backyard of her home in a Los Angeles barrio.

The Romero family at the funeral of an infant in Tucson, Arizona, in 1890.

LA FAMILIA

Ramiro Quintero's parents went to Texas after the Mexican Revolution, which had destroyed their estancia *(large farm). Born in the United States, Quintero recalled in 1989 the family life of his childhood.*

Our family's always been close. Comin' up, the biggest thing was discipline. The worst thing you could do was bring the folks *una queja* from a neighbor. A *queja* is basically a gripe or a complaint. "Your son broke a limb off my tree," or broke my window, or kicked the chicken. That would tear it. You'd get a whuppin' or a scoldin' or something. The way my dad would punish us is work. Every day he went to work he had chores for brothers or sisters to do....

If it's ten o'clock at night you'd finish it. There was always discipline and control over the family. That was the main thing, discipline and control. They were not whoopers or hollerers or screamers. They just spoke one time and that was it. The father was the breadwinner, the father was in charge of the whole house, what he said went, and the mother went along with him. There was no room for discussion. He was the boss and he never was wrong. No matter how wrong he could be, he still was right and that was the way of life. He kept real control of the family. He knew where everybody was at, and the later you came in at night the earlier you would get up the next day to go to work. You sure would, yes sir. Kept up on everyone, stayed on top of everything....

We didn't always have new clothes, but we had plenty to eat. How my father did it, I don't know. He was a horse trader. He'd trade chickens into the hogs. There might have been three or four hogs today an' he might have thirty or forty tomorrow. He learned the value of the dollar quick, and he understood the value of things.

Jacinta Carranza and her husband Simón Salazar Carranza had eight children. Simón worked for the railroad. As an old woman, Jacinta recalled her role as a mother.

I have never worked outside the home. My husband didn't want me to work—he said it wasn't necessary. He said that he supported me so that I could care for the children. And that is how I have spent my life—caring for my children. Even after my husband died in 1953, I did not leave the house to work. I worked right here—washing, ironing, cooking—and my children went to work to help out. I always loved my children very much—I never wanted to be apart from them because I worried that something might happen to them. They were always at my side.... I taught my daughters how to embroider, and I kept my sons sweeping and raking up the yard....

When my children were little I always used to tell them that they could not go to the movies if they did not go to Mass. So they would take their baths and get dressed and go to Mass and then I would let them go to the movies. I have always been religious and have had a lot of faith. I am especially devoted to the Virgen de Guadalupe, and she has been good to me and taken care of me and my children and protected us from harm. I prayed to her while my sons were in the Korean War and she brought them home safely.

Children were expected to show respect for their elders. Often, they were served their food after the adults had eaten. A Mexican woman from south Texas recalls an experience when, as a child, she complained:

Family portraits decorate the wall of this Mexican American farm family's home in Santa Maria, Texas.

All my relatives sat down to eat before us children. My mother had fixed chicken *mole* [spiced chicken stew], *frijoles,* rice, salad and fried potatoes. When the adults finished eating, she called us children to come in and eat. We sat down, and I realized that there was no fried potatoes left. That was my favorite dish. I made a fuss, complaining out loud that people had left nothing for us to eat. My mother almost died of embarrassment. She took the people to the other room and told me, "I am going to fix you now." Then she started peeling a mountain of potatoes right there and to fry them. She made a pile of fried potatoes and put them in front of me saying, "Now you are going to eat everything so that never again you complain about not having what to eat." I started eating but soon was full, and there was still a lot of potatoes to eat. I cried saying I couldn't eat anymore, and she simply said, "Go on eating, you are going to eat every single one of those potatoes." I begged my brothers to come help me eat, but she wouldn't let them do that either. Never again I complained about food in front of other people. My lesson was well taught.

Although the man was generally in charge in the Mexican home, sometimes there was a more equal sharing of authority. A woman in southern Texas looked back on the relationship of her parents, who lived on a ranch around 1910.

My mother *mandaba en la casa* [was the boss of the house], but decisions were made by both her and my father. If my father wanted to sell all the maíz or a cow, he first consulted with my mother. Of course, not all the couples around there were the same. There were some husbands who would lock the food in a compartment that only they could open. When the wife needed some food she'd have to ask him to open the compartment. A woman's life was very hard. We had little freedom and a lot of work. We had no horses, for instance, so we depended on the men for transportation. But it wasn't like that with my mama and papa. They were good to each other.

The Mexican American novelist Victor Villaseñor, in his book Rain of Gold, *told the history of three generations of his family, ending with his parents' marriage. In researching the book, Villaseñor interviewed many of his relatives.*

For myself, my biggest personal regret is that I never met my grandmother, Doña Margarita. She died two years before I was born. My father told me that he saw her only days before her death, shuffling down a dirt road in Corona, California, with the sunlight coming down on her through the tree branches. She was almost ninety years old, and he saw her walking along, doing a little quick-footed dance, singing about how happy she was because she'd tricked a little dog and he hadn't been able to bite her again.

My father said that tears came to his eyes, seeing how his mother—a little bundle of dried-out Indian bones—could bring such joy, such happiness, to her life over any little thing. "She was the richest human on earth, I tell you," said my father to me. "She knew the secret to living, and that secret is to be happy...happy no matter what, happy as the birds that sing in the treetops, happy as she came shuffling down that lonely dirt road, stopping now and then to do a little dance."

But...I did get to meet my mother's mother, Doña Guadalupe, and I was able to sit on her lap and have her rock me back and forth and tell me about the early days of La Lluvia when the gold had rained down the mountainsides and the wild lilies had filled the canyon with "heavenly fragrance."...

And I'm proud to say that I was able to finish the book before my father died. He was able to read it and see how I'd portrayed his loved ones, especially his mother. And on the last night of my father's life, I stayed with him, and his last words to me were, "I'm going to see *mi mama,* and I'm so proud of you, *mi hijito,* that you got her right in our book." He took my right hand in both of his, squeezing it, stroking it. "For she was a great woman," he said to me, "the greatest, just like your own mother!" And he hugged and kissed me goodbye.

I put him to bed, and he died in his sleep at the age of 86 or 84, depending on which relative I ask. All his life he'd been so strong and sure and confident and he died the same way. It wasn't that he had lost the will to live; no, he'd gained the will to die. For, I'd asked him, "Papa, aren't you afraid?"

"Of what?" he'd said in his deep, powerful voice. "Of death? Of course not. To fear death is to insult life!"

Elsie Chavez Chilton was born in Las Cruces, New Mexico, around 1915. Her parents had arrived from Mexico some years before. In 1983, Chilton recalled her childhood.

We had a big house. At that time, since the town was so sparsely populated, they built a house on each corner of the block so each block had only two houses. There was a vastness of brush, mesquite bushes, lizards and what have you. So we weren't crowded or close....

Rose Acevez Aguirre with her children David (the infant) and Blanca. The photograph was taken in a studio in San Gabriel, California, in 1911. The oceanfront scene is a painted backdrop.

We used to have a big dormitory room in our house for the five boys in our family. There were two of us girls. The boys would go out, I guess you would call it grubbing, for something to burn in the stove they had in their big room. They would get those mesquite roots, or tires or whatever else they could find and burn it in the big stove. We would have a roaring fire for a while and we were too hot. Then when the fire went out we would be cold again. Sometimes we would bundle up and go outside and play to keep warm. Mother would get up real early and go down to the *leniero* [the man who sold wood] for wood for the stove in the kitchen. For a quarter she would get a tub full of wood and then she would have a nice warm fire going by the time we got up to dress. When we got out of school the sun would be shining and it wouldn't be so cold.

The Spanish word compadre *can be translated as "close friend," but the relationship of a* compadre *is deeper than that. A* compadre *is a protector and guide. The most common way for a Mexican American to become a* compadre *is to sponsor a friend's child at its baptism. The person who does this is linked not only to the child but to its parents as well. One Mexican American explained the relationship in detail.*

A compadre means a lot. It's something real, it means a lot. When you make a compadre you have to respect him and he has to respect you. Compadres help each other; you can't talk about him, and he can't talk about you. For example, if you tell someone that your compadre is drinking too much then he may go over and tell your compadre that you were talking about him. Then your compadre will come to you and ask why you are talking about him. Then you may get into an argument and maybe you won't talk to each other after that. You shouldn't run around with the girls in front of him because of respect. You should try to show off that you're a nice man, and that you were chosen because you are a nice man.

Like you take Francisco, for example. He's a good friend of mine, but he wouldn't be good for a compadre. What I mean is that he comes into the house and jokes with me and my wife, he cusses around us, he doesn't respect us. He wouldn't be good for a compadre, but he's a good friend....

When you choose a compadre, you have to call him *Sir* in a way. You say *Usted* [the polite, respectful form of "you"]. When you see him on the street, you can't go rushing up to him and yelling, "Hey, you—come here!" If you know him real well, you address him by Sir.... Even if he is younger than you are, you address him nicely.

An elderly Mexican American man in a south Texas town recalled one of the traditional methods of courtship:

It used to be that when a boy's father went to the home of the girl to ask her father for his daughter's hand that the girl's father would say: "Well, I don't know. I'll tell you what. Send your boy over to my house for two months and we'll see how he works. We'll see what we can expect from that young man!" Then the father of the boy would agree to do this. But he would tell the father of the girl that he, also, wished to see how the girl behaved and what kind of a worker she was. He suggested that the girl come to his house for two months. That way there was an exchange [*intercambio*].

The girl would get up at three in the morning and see what kind of condition the *nixtamal* [tortilla dough] was in, and she would set the fire, bake the *tortillas,* and tidy up the place. At her family's home the boy would be working himself to death; he would be walking in from the fields with a load of sweet cane over one shoulder, and a load of corn on the other shoulder. People knew how to work and keep house then! After two months, when the parents had seen how well the children were able to work, they would give them permission to wed.

A woman makes tortillas in her home near Taos, New Mexico, in the 1930s.

Artemio Duarte, Jr., and his grandfather, Lauro Torres, in front of the family home on Del Amo Boulevard in Los Angeles. Artemio's father built the house himself.

Sometimes a young man would use a friend or other relative as a portador—an intermediary to approach the parents of the woman he wished to marry. One Mexican American in south Texas, returning from World War II, decided to marry his girlfriend.

She wanted to go off with me, but I...went and told my father that I wanted to get married and told him that I might *robar* [elope with] my bride. But he told me: "No, son; you must play it straight...." So that same week I got hold of two older men. I gave them each ten dollars.... They went to her house, and I went over to the Rincón bar to drink some beer. I was very nervous.

After a while they came over to the Rincón and sat down.... They said that the parents of the girl had told them to come back in eight days, and that's a sign that they approve of the marriage. If they had told them to come back in fifteen days for an answer, then it would have meant that they disapproved of the marriage.

I was very happy about the whole thing. In eight days I went over with my parents to make the first of the visits. While we were there, I told them that I was prepared to give her twenty dollars each week so that she could tell how I would support her. But she said, "No. Don't give me any money. Save the money and we'll go on a honeymoon instead." And that was all right with her parents; so when we got married we went to Monterrey and had a wonderful time!

Rose Reyes Pitts is a third-generation Mexican American. Her father's grandmother came to Edinburg, Texas, many years ago. Even so, as Reyes Pitts recalled, the spirit of machismo was still alive in her family.

My dad is very *mexicano*, very macho. In fact, even though I have a good relationship with him now, we never really communicated when I was growing up. It never bothered me, because I was so close to my mom. She always smoothed things out. I never realized it then, but now it sort of hurts my feelings. My father never really talked to me. If he wanted me to do something, he would tell my mother and my mother would tell me. When I married things got better; he had a little more respect for me; he talked to me. And since I've been a mother we've got along great.

My dad does all the men things. He barbecues, he keeps the yard perfect, keeps the cars. Of course he never changed a diaper. I can't even imagine that—I mean, I can't imagine being married to a man who would not help feed your child or change her diaper.

He always gives my mother a lot of credit though. He says she's the leader in the family, the one who's kept us all together, the smart one. And she is. She handles the household, pays all the bills, knows about the insurance and finances. He doesn't want to know about it. Even though he's real macho, she controls what happens, the important things. She really is in control of my father.

A family of Mexican immigrants in Houston, Texas, during the 1920s.

At a party in Tucson, Arizona, in 1989, a blind-folded girl swings at the piñata overhead. Filled with candy or toys, the piñata will release its prizes when broken.

A birthday party for Lulu Moreno in Los Angeles. She is the smallest girl, getting plenty of help blowing out the candles on her cake.

SCHOOL

A Mexican American schoolboy in a rural school in the Southwest during the 1930s.

In 1919, Lupe Valdez, who had come north with his parents after the Mexican Revolution, won a scholarship to the Texas-Mexican Institute, a school sponsored by the Presbyterian Church. Twenty years later, Valdez recalled his school days with fondness. But on one occasion, the teacher assigned the class a lesson about the Texas Revolution, in which the settlers won their independence from Mexico. Valdez and the other boys were angered to read about Sam Houston's "triumph" over the "cowardly" Mexicans. Valdez told an interviewer what happened.

That evening we called a meeting in one of the boys' rooms to talk the matter over. And the more we discussed it, the angrier we became at being assigned a lesson which showed the people of our blood in so unfavorable a light. What to do? First, in all of our copies of the book we defaced Sam Houston's picture beyond recognition. Then we decided the whole story had no place in a textbook from which we were supposed to study. This inspired one of the boys to rip the offending pages from his copy, and the rest of us, in a spirit of exultation, followed suit, telling each other we had paid for the books with the money provided...by our parents.

Next day an abnormally quiet group of boys filed into the classroom. The teacher asked us to turn to page so-and-so. My seatmate was the first to be called upon, and he replied with suppressed excitement, his black eyes dancing, that his book did not have that page. The teacher then called me. My copy, too, lacked the page. She called on the boy behind me, who announced that it was not in his book either. Puzzled, the teacher looked at our texts, then glanced at her own copy. Watching her, the class was so still you could hear the specks of dust dancing on the sheaf of sunbeams that poured in through a window. Finally, red as a beet, she nodded her head and said, "Let us take the next chapter for tomorrow. Class dismissed."

In many areas, Mexican Americans had to attend segregated schools. Even where the schools were integrated, Mexican Americans were not on an equal basis with other students, as one remembered.

We were the only six Mexicanos in my class. Most families could not send their kids to high school because it cost a lot of money, like for books, clothes, etc. We Mexicano students did not participate in anything like clubs, and were not represented in anything like being class officers, queens, sweethearts, you know, all that stuff kids like so much. Well, to belong to any club or to vote in the class, one had to pay a fee. That really was a problem. Our families were already making a big sacrifice by sending us to high school. We

The fourth-grade class at Drachman School in Tucson, Arizona, around 1913.

couldn't afford the extra costs. Besides, the gringos didn't make bones about showing their dislike for us. They snubbed us at all times. We were not wanted there. We were a minority. So we just stuck together and survived as best as we could. Many would just drop out because they couldn't bear the situation. It was my children's generation that started to change all this. Blessed they be.

Dolores Huerta grew up in Stockton, California, where her mother owned a restaurant and hotel. But in school she felt the sting of the teacher's prejudice.

When I was in high school I got straight A's in all of my compositions.... I used to be able to write really nice, poetry and everything. But the teacher told me at the end of the year that she couldn't give me an A because she knew that somebody was writing my papers for me. That really discouraged me, because I used to stay up all night and think and try to make every paper different, and try to put words in there that I thought were nice. Well, it just kind of crushed me....

Often Mexican children encountered teachers or counselors who believed they were suitable only for manual training classes. Robert Ortega, now an owner of a successful construction company, remembers:

My high school counselor advised that I was not college material, that I should do what my father did; that is, be a truck driver. What she meant was, "Stay where you're supposed to stay." She was Anglo. I was taught not to buck the system. It was a "yes, ma'am, yes sir," type of situation. What I did was go in and apply for the Naval Academy. And this was her response, "Why are you trying to leave here? Aren't you satisfied with your own life?"

It was kind of sweet justice later on in life: her husband was working indirectly under me. Things always come around.

Abel Vásquez, who later became a commissioner of Canyon County, Idaho, attended an integrated school there, but he sometimes felt out of place.

When I was in junior high school I wanted to go out for basketball. They said, everybody that wants to play basketball come on down. Needless to say, I was one of the first ones with all the other *gringitos* [white kids]. I looked like a *mosca* in the *leche* [a fly in a glass of milk]. In order to play you had to furnish your own sneakers. Well, I didn't have any. The first day, I was real good barefooted. The second day, I got blisters from playing on the hard wood. So I couldn't play basketball. My grandparents could not afford to buy me sneakers so I didn't participate because I didn't have shoes to play in.

Elsie Chavez Chilton recalled that when the Spanish-speaking children of Las Cruces, New Mexico, went to school in the 1920s, the teachers forbade them to speak Spanish:

The children would be out on the school yard and the minute the teacher would approach, why everybody would hush because they were speaking the only way they could—in Spanish.... I had already learned English from neighbors when we were in California so I was something of a novelty. The teachers used to keep me in at recess to sort of pump me. I guess they just wanted to hear a child speak English. They were tired of all the Spanish-speaking kids. Anyway, they would name us in English. They took away our Spanish names. My name, Elisa, became Elsie; my brother, Eduardo, became Edward or Eddie; and Samuel became Sammy and so on down. To this day we have our English names but my mother [then 99 years old] still calls us by our Spanish names.

Elsie Chavez as a child.

RELIGION

A woman prays in a church in a Mexican American neighborhood of Chicago.

In the late 19th century, Socorro Delgado's grandparents heard that a ranch was for sale near the town where they lived in Arizona. Delgado's grandmother gave her husband $200 that she had saved from the household funds. Then she prayed to Los Dulces Nombres *("The Sweet Names"—Jesus, Mary, and Joseph) that the ranch's owner would accept the money as a down payment. Her prayers were answered, and a century later, Socorro Delgado recalled how her grandmother would have a special ceremony every year on Christmas Eve to give thanks.*

She had a *velación* [vigil] in thanksgiving to Los Dulces Nombres because they had granted her prayer when they wanted to buy the ranch. She had the *velación* from the first year they moved to the ranch. My mother told me they would be a whole week in preparing for it. So Christmas was always special to her.... She always had this little altar, a little statue of Los Dulces Nombres which I still have and the two candles. But on Christmas Eve she would dress it up special. She covered the little table with beautiful crocheted and embroidered cloths that my great-grandmother had made. For many years I helped her decorate. We hung sheets on the wall and pinned a lot of paper flowers on them. She made the flowers herself.... And then she would buy four *big* candles at the cathedral and put two on each side of Los Dulces Nombres. I used to buy things at Kress' [store]—little birds and little bells—anything to hang to make it more colorful. And [her grandmother] would supervise and make sure that everything was done just so, that everything matched, that nothing was crooked. When we finished, she would say, "Tenemos que tener los claveles frescos." ["We must have fresh carnations."] She would always buy six fresh carnations to put on the altar. It was a *big* spending....

My parents, my sisters, Carmen and Andrea, and my brother, José, my uncles and aunts and cousins and neighbors and friends also came to my grandmother's house early on the twenty-fourth. Christmas Eve was a day of abstinence—you don't eat meat—so they made enchiladas and potato salad with egg. And she made biscochuelos (which are like cookies) and wine out of dried figs....

When people arrived they had a little snack, and then we went into her room to pray. The rosary was the important thing.... The waiting [for the actual *velación*] was till twelve, and then at midnight everybody went to see the Baby Jesus in the middle of the table; she had a cute little manger, and we went two by two, starting with the children, to adore the Baby Jesus.... Under the little altar table [grandmother] had hidden little gifts—a little doll or a little car or some candy from the dime store. She told us that the Baby Jesus had left it there because we had been good. We didn't know Christmas trees; we didn't know Santa Claus.

Virginia Gastelum recalled the religious devotion of her mother in Arizona in the 1920s.

My family was always very religious. My mother had an altar in her room. She liked to pray at night with candlelight or just one light. When she was finished praying, she would put all her prayer books away in a little tin box. She had all the saints on that altar! El Corazón de Jesus (Sacred Heart of Jesus), Santa Teresita, San Antonio, La Virgen de Guadalupe. The Virgen de Guadalupe was a picture in a frame as big as that wall! My mother always liked her altar to be clean. She had little doilies embroidered and with lace. She washed them and even starched them and put them on clean all the time.

We prayed the novena [nine days of prayer] to the Virgen de Guadalupe every year. My mother taught us that tradition. Every day we got together. Nine days of prayer—from December 4 to 12. And we'd stay all night on the vigil—the eleventh—to *amanecer* [greet the dawn]. We hung white sheets down the walls from the ceiling to the floor to make a little room just for the Virgin. We pinned flowers and little balls—like Christmas balls—all over. I used to climb ladders and stand on a table and sew them on with a needle and thread. We had lots of flowers, paper and plastic. And sometimes we had the fresh flowers, too, for La Virgen. We must have roses for the Virgin. Roses are her favorite flower, as you can see in her picture.

Socorro Delgado described how her grandmother achieved a lifelong wish—to make the pilgrimage to the cathedral in Mexico City where the Aztec Indian's cloak with the Virgin of Guadalupe's image is displayed.

My grandmother came here [to the United States] when she was three years old and all her life, because God had granted her so many blessings, she wanted to go visit the Virgen de Guadalupe in the Basilica in Mexico City. My cousins and I took her when she was seventy-five years old. She never got tired or sick the whole time. I have never seen anyone so happy, and the first time I ever saw her cry was when she saw la Virgen for the first time. When she knew we were going to take her, she went to Jacome's and bought herself a beautiful navy-blue dress out of georgette to wear on the day we went to the Basilica. She never wore that dress again, until twenty-two years later when she died. She kept everything she had worn on that trip—the stockings, underwear, and a little veil in a box on the top shelf of the wardrobe. She took it out from time to time to air it or send it to the cleaners. On the day of her funeral, something very odd happened; the funeral car that always went down Main to the cem-

As Others Saw Them

Adelina Otero Warren was superintendent of schools in Santa Fe, New Mexico, in 1929. In a horse-drawn wagon, she traveled to distant rural areas that were part of her district and wrote about the customs of the Mexican Americans living there:

Macario drives me over another mountain pass to Rio en Medio. On the top of this mountain...are more descansos [crosses] silhouetted against the blue sky. These are resting places where the people stop as they carry their dead to the grave. As we begin the ascent, I notice a procession going over the winding road. A wagon grinds its way ahead of a group of men. There are no women. The men are chanting. Macario tells me a young girl from Chupadero has died, and they are taking her to be buried in the church yard at Rio en Medio. The bell tolls in accompaniment to the chanting of the procession. *"Ese doble de campaña, no es por el que se murió, sino para recordarme a mí que me he de morir mañana"* (That tolling of the bell is not for the dead, but to remind me that I may die tomorrow), says Macario softly as he watches them, with hat in hand. The coffin was a home-made one, covered with pink calico, a white strip making a cross on its lid. A man is driving the team sitting on the box seat, while a young man kneels at the coffin's head, holding a cross. The rest of the men, about a dozen in all, walk bareheaded behind the wagon. At the top of the mountain they halt and a descanso cross is erected. The procession winds on, taking the high road. The chanting dies out among the trees as they disappear into the mountains.

After examining the school at Rio en Medio, I visited Macario's home. His wife embraced me, as is the custom among my people, and then invited me to have dinner which had been especially prepared in my honor.... They have a small adobe house of five rooms, with rag rugs on the floor, a spotless home. One room was set aside as a sanctuario, or little chapel, where candles were burning that day in front of images of the saints and of Our Lord. The family meet together every evening, light the candles, and say evening prayers. A devout, religious people whose religion permeates each thought and action. As these people watch the snow on the Sangre de Cristo range, they feel that it is a sign from Dios that there will be water in the streams for their crops. "Why are we Catholics? Because we are Spanish," some of our people say.

A Mexican American family on the occasion of the First Communion of the younger daughter.

etery took a turn and went around the block in front of Holy Family Church on West 3rd Street. Everyone was wondering: "*What* is he doing." "Está loco?" Later Leo Carillo (Sr.) told me that my grandmother had asked him to drive her by the front of Holy Family Church, because it was dedicated to Los Dulces Nombres, and he had remembered at the last minute.

Richard Rodriguez, a contemporary Mexican American writer, explained the importance of the Catholic church to his family.

I grew up Catholic at home and at school, in private and in public. My mother and father were deeply pious *católicos;* all my relatives were Catholics. At home, there were holy pictures on a wall of nearly every room, and a crucifix hung over my bed. My first twelve years as a student were spent in Catholic schools where I could look up to the front of the room and see a crucifix hanging over the clock....

My family turned to God not in guilt so much as in need. We prayed for favors and at desperate times. I prayed for help in finding a quarter I had lost on my way home. I prayed with my family at times of illness and when my father was temporarily out of a job. And when there was death in the family, we prayed.

I remember my family's religion, and I hear the whispering voices of women. For although men in my family went to church, women prayed most audibly. Whether by man or woman, however, God the Father was rarely addressed directly. There were intermediaries to carry one's petition to Him. My mother had her group of Mexican and South American saints and near-saints (persons moving toward canonization). She favored a black Brazilian priest who, she claimed, was especially efficacious. Above all mediators there was Mary, *Santa Maria,* the Mother. Whereas at school the primary mediator was Christ, at home that role was assumed by the Mexican Virgin, *Nuestra Señora de Guadalupe,* the focus of devotion and pride for Mexican Catholics. The Mexican Mary "honored our people," my mother would say. "She could have appeared to anyone in the whole world, but she appeared to a Mexican." Someone like us. And she appeared, I could see from her picture, as a young Indian maiden—dark just like me.

Rodriguez described the preparations for his First Confession and First Communion.

I n second grade, at the age of seven, we were considered by the Church to have reached the age of reason; we were supposed capable of distinguishing good from evil. We were able to sin; able to ask forgiveness for sin. In second grade, I was prepared for my first Confession, which took place on a Saturday morning in May. With all my classmates, I went to the unlit church where the nun led us through the forms of an "examination of conscience." Then, one by one— as we would be summoned to judgment after death—we entered the airless confessional. The next day—spotless souls—

we walked as a class up the aisle of church, the girls in white dresses and veils like small brides, the boys in white pants and white shirts. We walked to the altar rail where the idea of God assumed a shape and a taste.

As an eight-year-old Catholic, I learned the names and functions of all seven sacraments. I knew why the priest put glistening oil on my grandmother's forehead the night she died. At the baptismal font I watched a baby cry out as the priest trickled a few drops of cold water on his tiny red forehead. At ten I knew the meaning of the many ritual gestures the priest makes during the mass. I knew (by heart) the drama of feastdays and seasons—and could read the significance of changing altar cloth colors as the year slowly rounded.

David F. Gomez, who became a priest and Chicano activist, recalled the religious atmosphere in his parents' Los Angeles home in the 1950s.

The baptism of the newest member of a Mexican American family in Cicero, Illinois. The godparents, as the compadre *and* comadre, *will take a special interest in the child.*

The example of my parents had much to do with the nurturing of a vocation to the priesthood. In many ways they were typical Mexican Catholics. They always made sure that we children attended our catechism lessons and received our first communion and confirmation. My mother kept blessed religious statues around the house. In her bedroom was a statue of the Sacred Heart encased in glass with a votive candle burning in front of it. A statue of St. Martin de Porres was displayed in another room. Very popular among Mexican people, St. Martin was a seventeenth-century Dominican lay brother who was barred from the priesthood because he was a mulatto [of mixed white and black ancestry]; yet he went on to become one of the most popular saints in the Spanish-speaking world (perhaps because most Spanish-speaking people with mixed blood could sympathize and identify with him). My father's deep faith expressed itself in many ways. The good deeds and favors he did for our neighbors and relatives profoundly impressed me. He is the kindest, most generous man I have ever known.

Two Mexican American boys with the candle, rosary, and prayer book that are traditional gifts for a First Communion ceremony.

A procession at a church in Los Angeles. The image of the Virgin of Guadalupe remains a figure of deep reverence for the Mexican American community in the United States.

ASSOCIATIONS

The first Mexican American organizations were fraternal and mutual assistance groups. Carlotta Silvas Martin described the importance of such associations in Superior, Arizona, in the 1920s and 1930s.

There was a strong feeling of community among the Mexican people of Superior. The fraternal organizations had socials and picnics as well as dances. When someone died, they all turned out in their uniforms and banners—all very formal.

We also celebrated Las Fiestas Patrias—el 16 septiembre, el 5 de mayo, and also July 4. Some men in town knew a great deal about Mexican history and politics, and they were very involved in these celebrations. We'd have patriotic speeches, a queen and princesses, a parade with floats, a dance, a picnic with cakes and pie contests. On September 16 they'd blow up some dynamite, and the blast would represent "El Grito de Dolores," which was the cry for independence from Spain by Miguel Hidalgo.

We had a theater at one time in Mexican town. It had a very nice stage and seated about three hundred people. It burned down and was never rebuilt. Touring stock companies came from Los Angeles and presented beautiful plays in Spanish—comedies, dramas, and musical reviews. You know who used to come here—Rita Hayworth! [later a famous movie star] Her name was Rita Cansino then [she was a Mexican American]....

I belonged to a women's group called "Las Mexicanitas." We did a lot of community service. We raised money for scholarships; we donated the altar and railing when they built the new St. Francis Church and raised money for the St. Mary's Community Center. We put on plays and skits and variety shows. One year I wrote a play in Spanish and English called "The Apparition of Our Lady of Guadalupe." Martin Fierro, the high school Spanish teacher, did the men's voices, and I did the women's voices. My husband, Walter, and I went up to the falls and collected manzanita bushes. Walter put them on stands, and we decorated them with paper roses. Then we blacked out the stage lights, and when the lights were turned on the Virgen de Guadalupe was standing there. Everyone said, "Ahhhhhh." It was so beautiful. It was such a big success we had to put it on twice.

The first movie theater was in American town. Later we also had a movie house in Mexican town that had movies in Spanish. That's about the only time my mother would go out. How she enjoyed those movies!

In 1877, the officers of the Sociedad Patriótica de Juarez pose with a woman dressed for a play. The Sociedad was the most important civic organization in 19th-century Los Angeles, which in 1880 was still a small city of 11,000 people, mostly of Mexican descent.

In 1929, the League of United Latin American Citizens (LULAC) was formed to provide legal assistance and lobby for Mexican American rights. Felix Fraga, a resident of Houston, Texas, recalled the role of LULAC when he was a boy.

Before the war [World War II], even soon after, it was hard to mix Anglos and Mexicans here. All through the forties, you'd run across many signs that said, "No Mexicans allowed."

LULAC had to come in and take the school district to court, because they had schools where—no matter if you lived in front of it, you had to go to the Mexican school and not to the Anglo school. We went to our own parks and swimming pools, a little of what the black goes through, except it was harder for us, because at least they knew where they stood; we never knew. You'd go to a place and you'd wonder whether they'd tell you..."You can stay" or "You can't stay."

Ernesto Galarza described informal associations in the barrio.

The one institution we had that gave the *colonia* some kind of image was the *Comisión Honorífica,* a committee picked by the Mexican Consul in San Francisco to organize the celebration of the *Cinco de Mayo* and the Sixteenth of September, the anniversaries of the battle of Puebla and the beginning of our War of Independence. These were the two events that stirred everyone in the *barrio,* for what we were celebrating was not only the heroes of Mexico but also the feeling that we were still Mexicans ourselves. On these occasions there was a dance preceded by speeches and a concert....

Between celebrations neither the politicians uptown nor the *Comisión Honorífica* attended to the daily needs of the *barrio.* This was done by volunteers—the ones who knew enough English to interpret in court, on a visit to the doctor, a call at the county hospital, and who could help make out a postal money order. By the time I had finished third grade at the Lincoln School I was one of these volunteers. My services were not professional but they were free, except for the IOU's I accumulated from families who always thanked me with "God will pay you for it."

In 1929, an Anglo political boss in Texas was annoyed by the news that some of his Mexican American constituents had joined the newly formed local chapter of the League of United Latin American Citizens (LULAC). The boss scolded them in a letter to the local newspaper:

I have been and still consider myself as your Leader or Superior Chief.... I have always sheltered in my soul the most pure tenderness for the Mexican-Texas race and have watched over your interests to the best of my ability and knowledge.... Therefore I disapprove the political activity of groups which have no other object than to organize Mexican-Texas voters into political groups for guidance by other leaders.... I have been able to maintain the Democratic Party in power with the aid of my Mexican-Texas friends, and in all the time that has passed we have had no need for clubs or political organizations.

A patriotic parade organized by the Alianza Hispano Americana in Tucson, Arizona, around 1925.

During the Great Depression, a Mexican American union leader holds some food that a government agency distributed to needy families in San Antonio. These supplies—a sack of beans and some butter—were supposed to feed a family of three for two weeks.

In 1935, the federal government set up the Works Progress Administration (WPA), which created jobs by funding thousands of public-service projects. These Mexican American women in Minnesota attended an English class whose teacher was paid by the WPA.

DEPRESSION AND WAR

Elsie Chavez Chilton of Las Cruces, New Mexico, recalled the years of the Great Depression, when work was scarce. Her father had done construction work, but during the depression, few buildings were going up. He did not let that stop him.

My father was a dreamer. He was always treasure hunting. He resorted to placer mining in Orogrande [New Mexico] during the Depression.... He had a little lake and he put together a small conveyor and every once in a while they would hit a vein. I remember seeing really beautiful nuggets. He used to make tie pins. My daughter still has one of them. He would put little tiny bits of gold in some of the lockets. It was all very fascinating. He brought home a lot of turquoise. All our doors had doorstops of big old rocks of turquoise. At that time Indian jewelry and turquoise were not popular here.... I would keep little children's rings and bracelets to give at birthday parties. If a boy gave a girl a turquoise bracelet that was yuk, not good at all. We got by some of the bad years with my father doing the mining and jewelry making.

Despite their own hardships, the Chavez family was generous to those who were suffering even more. As Elsie Chavez Chilton recalled, people from Oklahoma who had lost their farms in the terrible drought of the 1930s went through Las Cruces on their way to California to find a better life.

Mesquite Street was a highway then and people traveling west from Oklahoma and other states would camp there when their vehicles broke down. Sometimes they camped there for weeks.... We would wind up helping them. They didn't have money to fix their cars.... My father was so generous he would help them out with gasoline, groceries, or whatever they needed and charge it to his account at the store. The lady at the post office and grocery store...saw everything. She would say, "Mr. Chavez, I'm going to keep that money and send it to your family. Don't be sending any more people over here to get gasoline or groceries on your account. The money you get from the gold that you mine doesn't get that far." That was a big help for our family when she took to handling my father's account.

My mother always used to say, *"Casa de herrero, azadon de palo,"* which means, "In the home of the blacksmith, wooden spoons." A long time ago, my mother's father was a blacksmith. They used to fashion the silverware also.

Socorro Delgado from Tucson, Arizona, described the effect of World War II on her community.

When the Second World War started, my brother José, my brother-in-law, my cousins, and my cousins' husbands served. At one time there were as many as fifteen of our immediate family who had gone to war! Every Friday Father Burns and Father Rossetti had a novena for the Sorrowful Mother. We would go and mention the names of the boys who were in the war. There was a victory candelabra with seven candles, and the candles would be lit. They lasted for a week. People came to my grandmother also and asked her to pray for their sons. Many were killed, but no one from our family was killed. My grandmother wanted to do something in thanksgiving for those who had come home safely, so she got together the families. Monseñor Timmermans gave us his blessing, and we made a procession to San Xavier Mission. I remember it was hot; we were all perspiring. We walked all the way! It was all open in those days; there were farms, milpas [cornfields]—nothing was fenced. I remember we had to cross acequias [irrigation canals] with running water. It's so different now! When we got to the mission we walked to El Cerrito where there was a grotto to the Blessed Mother. We prayed the rosary there in thanksgiving.

When the United States entered World War II in 1941, Mexicans responded by volunteering for the armed forces, where they served with distinction. Raul Morin, who wrote a history of the courageous deeds of Mexican Americans in two wars, explained why.

Even with the constant discrimination and continued denial of equal opportunities, when war came to the United States, no one could accuse us of draft dodging or fleeing to Mexico to avoid military service as [was] charged in World War I.

Just prior to Pearl Harbor [the Japanese attack on a Hawaiian naval base that spurred the United States to declare

As Others Saw Them

During the early months of World War II, U.S. troops fought to defend the Philippine island of Bataan against Japanese attacks. About one-fourth of the U.S. forces on Bataan were Mexican Americans. After the war, their commander, General J. M. Wainwright, recalled the valor of the Mexican American troops:

It seems to me that the rediscovery of the unusual devotion to duty, the courage and the willingness to sacrifice for the good of the country of all kinds of Americans was one of the lessons of World War II.

Most of us came to understand that membership in a minority group did not set us apart from our neighbor because we are all members of one form of minority or another.... In combat we asked no questions about a man's background so long as we knew him to be a soldier.

The part played in World War II by men and women of Mexican extraction was considerable. Almost every unit in the United States Army included Mexican-American soldiers and they served well.

Just recently it has been my privilege to present...the Silver Star, the Bronze Star, and the Purple Heart to the parents of a Mexican boy from Texas who gave his life in the second battle for the Philippines. Within the past year I have presented the Medal of Honor to three American soldiers of Mexican parentage.

Examples such as this are numerous. Anyone would be proud to have served in the same Army with these men.

The men in uniform are Sergio and George Moriera, two of the nearly 500,000 Mexican Americans who served in the armed forces during World War II. After the war, Mexican American families in Los Angeles erected a memorial to their war dead.

José P. Martínez

On May 27, 1943, the War Department of the United States reported that on the previous day Private José P. Martínez had been killed in action. The young Mexican American had died a hero while fighting Japanese troops. He became the first draftee in World War II to be awarded the Congressional Medal of Honor.

José Martínez was born 23 years earlier in Colorado. Working as a farm laborer, he was known as a mild-mannered, likable person to his friends. With the outbreak of World War II, Martínez was drafted and eventually sent to the Aleutian Islands that lie between Alaska and Russia.

In the days after the attack on Pearl Harbor, the Japanese also seized Attu in the Aleutians. Eighteen months later, Martínez and his platoon were part of the mission to eject the well-entrenched Japanese.

The U.S. attack began on the morning of May 26, 1943. The fighting took place in snow-covered mountains. Though Martínez's battalion succeeded in landing on the island, they were soon pinned down by Japanese machine-gun, rifle, and mortar fire. Then Martínez stepped forward. With bullets whizzing around him, he charged forward alone. Turning to his comrades, Martínez encouraged them to follow. Amazingly, the men of the whole battalion began to follow a lowly private. After capturing two Japanese trenches, Martínez was struck down, still fighting.

Part of his citation for bravery read: "He was mortally wounded with his rifle still at his shoulder, absorbing all enemy fire and permitting all units to move up behind him and sucessfully take the pass."

Word of Martinez's exploits spread. He became the special pride of the Mexican community at a time when Mexican Americans in Los Angeles were facing attacks from soldiers and sailors in the zoot suit riots.

war], many developments had occurred among the Spanish-speaking living in the USA. We felt that a great change was being made in our lives since the dark days of the Depression. A change that was to transform us from the lackadaisical, timid person to an alert, progressive, up-to-date citizen. The feeling was that little by little we had been forsaking the old way of doing things, we had learned more of the American way of life; and we could foresee the day when we Americans of Mexican descent would be fully integrated with the vast number of other ethnic groups that make up America.

Most of us were more than glad to be given the opportunity to serve in the war. We knew there was something great about this country that was worth fighting for. We felt that this was an opportunity to show the rest of the nation that we *too* were also ready, willing, and able to fight for our nation. It did not matter whether we were looked upon as Mexicans, Mexican-American, or belonging to a minority group; the war soon made us all *genuine* Americans, eligible and available immediately to fight and to defend our country, the United States of America....

Our people have the same common spirit. Here were those individuals who had never met before, and yet at the very first contact everyone acted as though they had known each other for a long time.

On the first night at the reception center, all the gang, approximately 30, as if by a prearranged plan, automatically gathered outside the barracks to get better acquainted. Soon someone brought out a guitar and we began singing songs, old and new, all familiar to us.... The songfest kept getting louder and the tempo picking up....

Army life afforded many their first experience in putting into practice a real democracy, a simple, quiet democracy effected without any pressure or compromise. We were so engrossed in our new chores of soldier life and so aware of an uncertain future, that no room was left for anyone to be choosy about his neighborhood.

The war gave many women their first opportunity to work outside the home. Margarita Avila, who was in her early twenties, found her first job after moving to Los Angeles. She recalled the thrill it gave her.

It was the time of World War II, the end of the war, really. Seeing so many movies, I thought of myself as a heroine. I wanted to work in the war effort, in an airplane factory or something like that. I wound up working in a sewing machine factory. They asked if I had any experience, and I said, "Yes." They asked if I could come in tomorrow, and I said, "No, I want to start right now."

Well, I start to work. They seat me at a machine, and I have never worked a sewing machine. A woman who worked next to me told me it was run by your knee. It sounded easy. When I sat down, the boss was standing right behind me. I started leaning forward and I had to put my feet on the pedal, and the machine went off, Ruh-h-h-h! I got so frightened I went over backward in my chair. Of course, the boss saw all this, and started to laugh. He joked, "Experienced?" I answered, very annoyed, "Yes."

I worked there for about a year. During that time I got married and got pregnant. Sometimes I would go in for a month to work, sometimes even for a day, but I didn't work regularly anymore after my son was born.

Two Mexican American servicemen pose with their girlfriends during World War II. Mexican Americans who served in the war returned home with a determination to assert their full rights as U.S. citizens.

During World War II, many women of all groups left their traditional roles as homemakers and took the jobs of men who had gone to war. Most of these railroad workers photographed in Arizona during the war were Mexican American.

One of the important celebrations in the life of a Mexican American woman is her quinceañeras, or 15th birthday. At her party, this young woman poses in front of a mural of pop stars.

CHAPTER SIX

PART OF THE UNITED STATES

aving distinguished themselves for their bravery in World War II, Mexican Americans came home determined to demand equal rights as citizens. Moreover, Mexican American women, who contributed to the war effort at home, gained greater economic independence and learned from their involvement in labor unions how to organize to make their demands felt.

However, they still faced the prejudice and discrimination that had held them back in the past. In 1948, a Texas funeral parlor refused to bury Félix Longoria, a Mexican American veteran who had been decorated for heroism, in an Anglo cemetery. Anger over this incident inspired Dr. Hector García of Corpus Christi to found the American G.I. Forum. By purposely omitting the word *Mexican* from its title, the organization's leaders made clear their goal of claiming full rights of American citizenship. The American G.I. Forum rapidly gained members, and it played a major role in eliminating discrimination in the workplace and in public facilities such as schools and theaters.

Other organizations that sprang up in the years after World War II brought Mexican Americans into politics. The Unity Leagues, founded by Ignacio López, organized voter registration drives and supported the candidacies of Mexican Americans running for local offices, primarily in California. Such groups as the Mexican-American Political Association (MAPA) in California, the Political Association of Spanish Speaking Organizations (PASSO) in Texas, and the American Coordinating Council for Political Education (ACCPE) in Arizona did similar work. Older organizations such as LULAC and the Alianza Hispano Americana also remained active.

Their efforts began to pay off in 1949 with the election of Edward Roybal to the Los Angeles City Council. He was the first Mexican American to serve on the council since 1881. Seven years later, Henry B. González became the first Mexican American member of the Texas State Senate in 110 years. Later, both González and Roybal won seats in the U.S. House of Representatives, where they have served with distinction.

Though political action groups achieved some success, they found it hard to reach the Mexican Americans at the bottom of the economic ladder—the migrant workers. Because they moved from place to place during the year and because many of them were illegal immigrants, the migrants were reluctant to antagonize the growers who exploited them.

Several farm workers' unions, mainly in California, were founded in the late 1920s and early 1930s, but the Great Depression and the repatriation program hampered their efforts. After World War II, there were further attempts to organize the farm workers, but they had little success. Then in 1962, César Chávez started the National Farm Workers Association (NFWA). Using his life savings, Chávez drove to the migrant camps in his old car, slowly attracting members to the NFWA. Three years later, in 1965, Chávez's group merged with an organization of Filipino workers to form the United Farm Workers (UFW). The UFW's members voted to go on strike against growers of table grapes around Delano, California.

La Huelga, "the strike," grew from a local affair into a movement that brought Chávez and the UFW national attention. Finally, in 1970, the growers signed contracts with the UFW.

The triumph of *La Huelga* was a source of pride for Mexican Americans. It coincided with the beginning of what became known as the Chicano movement. *Chicano* was a commonly used slang word referring to the very poor, but in the United States in the late 1960s, young Mexican Americans adopted it as a mark of self-identification and pride.

Chicano activists sometimes adapted the slogans and techniques of the African American rights movement. "Brown power" was a rallying cry of the Brown Berets, a student-led group that organized a walkout of 10,000 Chicanos from East Los Angeles high schools in March 1968 to protest substandard conditions in barrio high schools. Young people of barrios in California, Arizona, Texas, and Colorado organized demonstrations to protest discrimination in schools, police brutality, and the racism that relegated them to menial, low-paying jobs.

Chicanos also protested U.S. involvement in the Vietnam War. Because college students could obtain deferments from military service, the U.S. combat troops were largely composed of young men who had been denied the opportunity to attend college. (In 1971, only about 1 out of every 20 Mexican Americans was a college graduate, compared with 1 out of 4 Anglos.)

Chicano leaders declared August 29, 1970, as National Chicano Moratorium Day, when antiwar demonstrations would be held throughout the nation. A huge crowd assembled in the East Los Angeles barrio and marched peacefully toward Laguna Park. Police began to fire tear gas into the crowd, and a full-scale riot broke out. During the violence that followed, police shot a tear-gas canister into a bar, striking and killing Rubén Salazar, who was sitting at a table. No one was ever indicted for his murder.

Today's Mexican Americans cherish and seek to preserve their cultural roots. Here, in Chicago in the 1980s, a couple performs the Mexican hat dance.

Rubén Salazar was a prominent Mexican American newspaper and television journalist. He had written articles that helped to define the Chicano movement, declaring that it was more than just a protest against decades of injustice. *Chicanismo* represented Mexican Americans' pride in their culture.

The literary and cultural aspects of the Chicano movement blossomed. Rodolfo "Corky" Gonzales, a former pro boxer, wrote an epic poem titled *I Am Joaquín*, which was passed through the barrios, appeared in book form in 1967, and was made into a movie. Many other Chicano writers began to publish their essays, fiction, and poetry. Artists painted large murals, like those that had been created in Mexico earlier in the century, to celebrate the history of Mexicans and Mexican Americans. Theater groups presented new plays that glorified the Chicano heritage.

Some Chicanos adopted the ideal of Aztlán, the legendary homeland of the Aztecs, to assert their ancient roots in the southwestern United States. As David F. Gomez, a Catholic priest and Chicano activist, wrote in 1973, "Aztlán will be our own independent, free nation where we will no longer be the colonized, exploited, and abused people we have been for generations."

The movement extended into political action with La Raza Unida party, begun in Crystal City, Texas, in the early 1970s. Mexican American high school students there boycotted classes after the selection of an all-Anglo cheerleading squad. José Angel Gutiérrez formed the La Raza Unida party to back Chicano candidates for the city's school board. After they won, the schools adopted Chicano studies and bilingual education. Other La Raza Unida branches were organized throughout the southwestern states, electing candidates to local offices.

Though the party lost its original impetus, other groups such as the Southwest Voter Registration Education Project have continued to help Mexican Americans exercise their political influence. From 1974 to 1984, the Southwest Voter Project carried out more than 1,150 voter registration drives in Arizona, New Mexico, California, Colorado, and Texas. In 1974, two Chicanos, Jerry Apodaca in New Mexico and Raul Castro in Arizona, were elected governors of their states. Since then, many Mexican Americans have been elected to public office.

In addition, the Mexican American Legal Defense and Education Fund (MALDEF), founded in 1968, has fought legal battles against discrimination and segregation and has sponsored the education of Mexican American lawyers.

One issue that still remains controversial is bilingual education. Mexican Americans, along with other Spanish-speaking groups, have insisted that their children be taught in the language they speak at home. The Supreme Court agreed that it was a violation of students' rights to teach them in a language they did not fully understand. In 1965, Congress passed the Bilingual Education Act, providing funds to encourage teaching in Spanish. However, during the 1980s, federal funding for bilingual programs was cut. Sixteen states have passed laws against some forms of bilingualism.

Nevertheless, many school systems now teach children in both English and Spanish, recognizing that it is often the most effective way to educate those who are not proficient in English.

The burst of enthusiasm that marked the early years of the Chicano movement did not endure. However, its effects were longer

Public housing projects in the 1960s gave many poor people decent places to live in urban areas throughout the United States. These boys lived in the Ramona Gardens project of Los Angeles.

lasting. Mexican Americans have persisted in their efforts to take their rightful place in U.S. society. Many universities now have Chicano studies departments, and textbooks reflect a more accurate picture of the relationship between the United States and Mexico.

The publication in 1972 of Rodolfo A. Anaya's novel, *Bless Me, Ultima,* awakened interest in the writings of Mexican Americans. Since then, writers such as Sandra Cisneros, Ana Castillo, and Gary Soto have won recognition.

Mexican Americans have also gained fame in show business. In 1993, a brilliant young movie director, Robert Rodriguez, made his debut with the film *El Mariachi,* which he made on a tiny budget of $7,000. Talented Mexican American actors such as Edward James Olmos and Ricardo Montalban have attained star status in movies and television. Linda Ronstadt and the popular band Los Lobos have brought Mexican musical styles into the mainstream of United States pop music.

But perhaps the best-known symbol of Mexican culture is the fiesta. Today some of the great Mexican American celebrations center on such national events as Mexican Independence Day and Cinco de Mayo, as well as saints' day celebrations. The fiestas show the pride that Mexican Americans feel for their roots and demonstrate that their culture has endured as part of U.S. civilization.

Mexican Americans have also brought their cooking to their Anglo neighbors throughout the United States. Chili is perhaps the most popular of all Mexican American dishes, but tacos, enchiladas, and tortillas are becoming almost as familiar. Taco Bell operates more than 4,000 restaurants that serve over 2 billion tortillas a year. Salsa, a spicy tomato-based sauce, rivals ketchup in supermarket sales. Aztec women, who spent hours grinding corn kernels to make the tortillas that were their families' staple food, would be surprised to see how that tradition has developed in the United States.

César Chávez

"The fight is never about grapes or lettuce," César Chávez said. "It is always about people." Chávez was one of the United States's most important labor and civil rights leaders. La Causa, the popular name for the movement he inspired, united many Mexican Americans and gave them a new pride in their heritage.

César Chávez was born near Yuma, Arizona, in 1927. His grandfather had come from Chihuahua, Mexico, in the 1880s. The family lost their farm during the Great Depression, and the Chávezes turned to migrant farm labor in California. Life was hard during César's childhood. He would never forget the prejudice of segregated schools and the look on his father's face when he was refused service in a luncheonette.

After serving in the navy in World War II, Chávez settled in San Jose with his new wife, Helen. Their home was in the barrio known as "Sal Si Puedes" or "Get Out If You Can." In 1952, Chávez joined the Community Service Organization (CSO), an association dedicated to helping people in poor neighborhoods.

Ten years later, Chávez decided to organize the farm workers into a union. Traveling in his old car, he enlisted members in the fields and migrant worker camps. Drawing on their heritage, they took the Aztec eagle as their symbol.

In 1965, Chávez's union decided to join 800 Filipino grape pickers who had gone on strike. *La Huelga*—the strike—would last five years.

To win public support for the strike, Chávez led his members on a 300-mile march to the state capital. When the growers used violence against picket lines of Chávez's United Farm Workers (UFW), he began a 25-day fast in protest. He called for a nationwide boycott of table grapes to pressure the growers to negotiate.

The boycott won nationwide support, and by 1970, most of the growers had agreed to the workers' demands, increasing their wages from $1.10 to $2.10 an hour.

César Chávez would go on to organize the lettuce workers as well. In the 1980s, however, the membership of the UFW declined and many of the gains that had been made in the 1960s were lost. A hostile administration in Washington, as well as mistakes made by Chávez's organization, accounted for the failures. Yet when César Chávez died in 1993, more than 35,000 people came to pay respects to the man who had done so much for his people.

LA HUELGA

Through sheer determination, César Chávez created a labor movement among the migrant farm workers who had previously enjoyed virtually no rights. He described the beginning of the United Farm Workers in Delano, California.

A lot of people have asked me—why Delano, and the answer is simple. I had no money. My wife's family lived there, and I have a brother. And I thought if things go very bad we can always go and have a meal there. Any place in the Valley would have made no difference.

I had some ideas on what should be done. No great plans; just that it would take an awful lot of work and also that it was a gamble. If I can't organize them to a point where they can carry on their own group then I'm finished, I can't do it, I'd move on and do something else.

I went around for about 11 months, and I went to about 87 communities and labor camps and in each place I'd find a few people who were committed to doing something; something had happened in their lives and they were ready for it. So we went around to the town, played the percentages, and came off with a group.

We had a convention here in Fresno, the first membership meeting, to set up a union—about 230 people from as many as 65 places. We knew the hardest thing would be to put across a program that would make them want to pay the $3.50 (monthly dues), because we were dependent on that. I felt that organizing couldn't be done on outside money.

We had signed up about 1,100 people. The first month 211 paid. At the end of three months we had 10 people paying. Talk about being scared! But we went back and kept at it. By this time Dolores [Huerta] was helping me up in the Northern part of the Valley, and I was getting help from Gilbert Pedilla, both of whom are Vice-Presidents now. Gradually the membership was increasing. At the end of six months we were up to about 200 members. Instead of going all over the Valley as I did at first, I started staying one place long enough for them to get in touch with me if they wanted to. We put a lot of emphasis on the people getting members.

We had hundreds of house meetings. Sometimes 2 or 3 would come, sometimes none. Sometimes even the family that called the house meeting would not be there.

I was trying to prove anything to anyone who had given money. If I'd been under a board or a grant I don't think it would have worked. In the first place, I had to get the dues in order to eat. I suspect some of the members were paying dues because they felt sorry for me.

Dolores Huerta gave up a well-paying job to become an organizer for the United Farm Workers union. She later confessed that her father and other family members criticized her for devoting too much time to her organizing work and not enough to her family. Huerta told an interviewer:

If you let it bug you when people say that you're not being a good mother because you're not with your kids twenty-four hours a day, well then of course it will deter you from what you're doing. In the union, you know, everybody cooperates to take care of your kids.

The idea of the communal family is not new and progressive. It's really kind of old-fashioned. Remember when you were little you always had your uncles, your aunts, your grandmother, and your comrades around. As a child in the Mexican culture you identified with a lot of people, not just your mother and father like they do in the middle-class homes. When people are poor their main interest is family relationships. A baptism or a wedding is a big thing....

While I was in jail some of my kids came down to Delano to see me, but my little girl, Angela, didn't come. She wrote me a little note which said, "Dear Mom. I love you very much, but I can't come because the people need me. I've got to go door-knocking this weekend and I can't leave my job." I think that's really great because she puts her priorities on the work she has to do instead of coming down to see me.

Women played a prominent role in the UFW union. Jessie Lopez De La Cruz was the daughter of a migrant farm worker family who lived in the San Joaquin Valley. She describes her experiences as an organizer.

Growing up, I could see all the injustices and I would think, "If only I could do something about it! If only there was somebody who could do something about it!" That was always in the back of my mind. And after I was married, I cared about what was going on, but felt I couldn't do anything. So I went to work and I came home to clean the house, and I fixed the food for the next day, took care of the children and the next day went back to work. The whole thing over and over again. Politics to me was something foreign, something I didn't know about....

But then late one night in 1962, there was a knock at the door and there were three men. One of them was César Chávez. And the next thing I knew, they were sitting around our table talking about a union. I made coffee. Arnold had already told me about a union for the farmworkers. He was attending their meetings at Fresno, but I didn't. I'd either stay home or stay outside in the car. But then César said, "The women have to be involved. They're the ones working out in the fields with their husbands. If you can take the women out to the fields, you can certainly take them to meetings." So I sat up straight and said to myself, "*That's* what I want!"

When I became involved with the union, I felt I had to get

César Chávez in a demonstration for a group of California farm workers around 1968. By devoting his life to La Causa, he won nationwide respect for himself and for all Mexican Americans.

A woman carries the UFW banner with the Aztec eagle.

101

Luis Valdez

In 1965, Luis Valdez headed for Delano, California, to support the strike by the United Farm Workers. He also went to realize a personal dream—the creation of the Teatro Campesino, a theater group made up of the strikers themselves. At Delano, Valdez later recalled, there was "no money, no props, no scripts, no stage, no lights, no actors, nothing. All that was there was just the spirit of the people, but that was enough."

Luis Valdez's parents were migrant workers, which meant that as a child, he traveled from school to school when the work season demanded. In first grade, he was to be in a play for which the children had made papier-mâché masks. Young Luis was thrilled with his role, but the family had to move on before he could perform. Still, the experience inspired in Valdez a lifelong love for the theater. As a child, he put on his own plays, making puppets and using a large cardboard box for the stage.

Despite the migrant life, Valdez graduated from high school and received a scholarship to San Jose State College. He went on to become part of the San Francisco Mime Troupe until he heard about the farm workers' strike. Soon, his Teatro Campesino was raising money for the farm workers' cause by performing on college campuses.

Two years later, Valdez left the fields to develop his art. He wrote short plays called *actos*, in which he explored themes related to the Chicano movement. Often, he used folklore and elements of Aztec culture. In the 1970s, Valdez wrote *Zoot Suit*, a play based on the 1943 riots in Los Angeles. The play was such a success that it later appeared on Broadway and was made into a motion picture.

Valdez turned his artistic skills to the screen, writing and directing the movie *La Bamba*. Originally a dance from Mexico, "La Bamba" was the title of a hit record by a Chicano rock star of the 1950s who died in a plane crash. The movie was the life story of that young Chicano, Ritchie Valens (born Ricardo Valenzuela). It became a big hit among both Spanish- and English-speaking audiences.

other women involved. Women have been behind men all the time, always. Just waiting to see what the men decide to do, and tell us what to do…. I'd hear [women] scolding their kids and fighting their husbands and I'd say, "Gosh! Why don't you go after the people that have you living like this? Why don't you go after the growers that have you tired from working out in the field at low wages and keep us poor all the time? Let's go after them! *They're* the cause of our misery!"…

It was very hard being a woman organizer. Many of our people my age and older were raised with the old customs in Mexico: where the husband rules, he is a king of his house. The wife obeys, and the children, too. So when we first started it was very, very hard. Men gave us the most trouble…. They were for the union, but they were not taking orders from women, they said….

That year, we'd have a union meeting every week. Men, women, and children would come. Women would ask questions and the men would just stand back. I guess they'd say to themselves, "I'll wait for someone to say something before I do." The women were more aggressive than the men. And I'd get up and say, "Let's go on, let's do it."

Roberto Acuna's parents were farm workers, and he spent his own childhood working in the fields. When he was around 30 years old, Acuna became an organizer for the United Farm Workers. He told an interviewer that at first he resisted the union.

I would read all these things in the papers about César Chávez and I would denounce him because I still had that thing about becoming a first-class patriotic citizen. In Mexicali they [union members] would pass out leaflets and I would throw 'em away. I never participated. The grape boycott didn't affect me much because I was in lettuce. It wasn't until Chavez came to Salinas, where I was working in the fields, that I saw what a beautiful man he was. I went to this rally, I still intended to stay with the company. But something—I don't know—I was close to the workers. They couldn't speak English and wanted me to be their spokesman in favor of going on strike. I don't know—I just got caught up with it all, the beautiful feeling of solidarity.

You'd see the people on the picket lines at four in the morning, at the camp fires, heating up beans and coffee and tortillas. It gave me a sense of belonging. These were my own people and they wanted change. I knew this is what I was looking for. I just didn't know it before.

My mom had always wanted me to better myself. I wanted to better myself because of her. Now when the strikes started, I told her I was going to join the union and the whole movement. I told her I was going to work without pay. She said she was proud of me. (*His eyes glisten. A long, long pause.*) See, I told her I wanted to be with my people…. I had to belong to somebody and this was it right here. She said, "I pushed you in your early years to try to better yourself and get a social position. But I see that's not the answer. I know I'll be proud of you."...

Working in the fields is not in itself a degrading job. It's hard, but if you're given regular hours, better pay, decent housing, unemployment and medical compensation, pension plans—we have a very relaxed way of living. But the growers don't recognize us as persons. That's the worst thing, the way they treat you. Like we have no brains. Now we see they have no brains. They have only a wallet in their head. The more you squeeze it, the more they cry out.

If we had proper compensation we wouldn't have to be working seventeen hours a day and following the crops. We could stay in one area and it would give us roots. Being a migrant, it tears the family apart. You get in debt. You leave the area penniless. The children are the ones hurt the most. They go to school three months in one place and then on to another. No sooner do they make friends, they are uprooted again. Right here, your childhood is taken away....

When people have melons or cucumbers or carrots or lettuce, they don't know how they got on their table and the consequences to the people who picked it. If I had enough money, I would take busloads of people out to the fields and into the labor camps. Then they'd know how that fine salad got on their table.

Luis Valdez, who formed the Teatro Campesino to entertain and educate the striking farm workers in Delano, California, describes the effect of the strike, which continued for five years.

After years of isolation in the barrios of Great Valley slum towns like Delano, after years of living in labor camps and ranches at the mercy and caprice of growers and contractors, the Mexican-American farmworker is developing his own ideas about living in the United States. He wants to be equal with all the working men of the nation, and he does not mean by the standard middle-class route. We are repelled by the human disintegration of peoples and culture as they fall apart in the Great Gringo Melting Pot, and we are determined that this will not happen to us.

The success of the United Farm Workers brought great pride to many Mexican Americans. One supporter, who witnessed the three-week march from Delano to Sacramento, declared:

Before 1965 I can say that we lived in the dark ages because we were not involved as a people, not socially aware of the problems and injustices.... [Then] César Chávez...marched from Delano to Sacramento. The marchers arrived on Easter Sunday and there were ten thousand people. That was the largest group of Mexican-Americans I had ever seen. It was a beautiful experience.... From then on there was a new spirit among the Mexican-Americans, a new spirit of awareness and concern.

Photographs of farm worker demonstrations touched the hearts of many other Americans. The widespread support given to the grape and lettuce boycotts helped the migrant laborers win their demands for better pay and living conditions.

THE CHICANO MOVEMENT

La Raza Unida party members in Houston during the 1972 elections. The party gave many young Mexican Americans their first experience with politics. Some continued to work for change after the party itself had declined.

Social worker Lydia Aguirre described the Chicano movement's aim.

Chicano power simply means that in the finding of identity—that is, a right to be *as he is,* not Mexican, not Spanish, not speaking either a "pure" English or a "pure" Spanish, but *as he is,* a product of a Spanish-Mexican-Indian heritage and an Anglo-Saxon (American...) influence—he will unite with his brothers in heritage.

We demand not to be segregated. We demand that others recognize our differentness and work within that differentness rather than make the Chicano suppress his Chicanismo and adopt Anglo-Saxon ideals.

Some Chicanos...attempt to effect a change within established systems, and if that does not work, attempt to establish separate systems.... Other Chicanos prefer isolationism or brown separatism. These are in the minority. Very few would want a separate nation.

The term Chicano *means different things to different people. In 1972, the* DQU Newsletter *of D–Q University in Davis, California, described it in an article entitled "Are You a Chicano?"*

The word "Chicano" has been used among the Mexican people for years, with both a good and bad interpretation of the term. Recently, the newspaper and television medias have been using Chicano as a method of identifying our people and our movement.

Many of us prefer to call ourselves Chicanos because it reflects our feeling for our culture and heritage, our pride in who we are and where we came from, and it gives us a sense of unity. But that's not all the term means; besides unity it has brought suspicion and has tended to separate the "Chicanos" from many others of Mexican descent through a misunderstanding of the term, the way it is used, and each other.

Lately, Chicano has been used as a kind of measuring stick. This has served to split us into groups based solely on what we choose to call ourselves. That we are wasting our time and energy arguing among ourselves about what name we should go under is just one of the problems raised by this "hang-up" of ours. What difference does it make?

Some of the time it seems that many of us forget that we are a very diverse group of *individuals.* Many of us don't know the field, the barrios, the Spanish language, or even the pronunciation of our own names. This might not be the way we would like it to be, but that's the way it is. These things are nobody's fault, they just happened and we should be ready to go from there....

The major "virtue" that we lack is patience. Apparently it's been too long for some of us to remember that all of us didn't always have the pride in La Raza we have now; we didn't always know the difference between doing something for ourselves and doing things for our people; and all of us didn't grow up calling ourselves Chicano. It's time that we brought each other along, and took the time to explain to each other our pride and our feelings about what we're doing.

The problems of education, housing, unemployment, etc. are still there waiting to be solved by Chicanos, Mexicans, Mexican-Americans, Latinos—in short, anyone who sincerely wants to do the work. So it's not a matter of what you call yourself—if you want to help, DO IT! And you'll be whatever you want to be, and be called whatever you want to be called.

Rodolfo "Corky" Gonzales, one of the most important Chicano leaders, reflected on the special difficulties of being a Mexican American in his epic poem called I Am Joaquín. *The poem's title refers to the legendary Mexican American bandit Joaquín Murieta. But Gonzales's subject matter includes all of Mexican and Mexican American history. The poem begins:*

> I am Joaquín
> in a country that has wiped out
> all my history,
> stifled all my pride.
> My knees are caked with mud.
> My hands are calloused from the hoe.
> I have made the Anglo rich.
> Here I stand
> Poor in money
> Arrogant with pride.

Texas-born Reies López Tijerina, a Protestant minister, was one of the most charismatic leaders of the Chicano movement. In 1966, López Tijerina founded the Alianza de los Pueblos Libres (Alliance of Free City-States), which sought to reclaim old Spanish and Mexican land grants in what is now the state of New Mexico. In June 1967, López Tijerina led a group of followers on a raid of the courthouse at Tierra Amarilla, New Mexico. After he was taken into custody, López Tijerina made the following statement.

We are more determined than ever to win back our land, to preserve and revive our culture.... Circumstances themselves are forcing la raza together. We will have a movement all across the southwest; nothing can stop it. And our people are forced by the same circumstances, and the same fight, irresistibly towards the Negro people and their fight. This urge, this growth, for unity of la raza with the Negro people cannot be resisted, it will come because it must.... We, la raza, are a new people, a young people, only 300 years old—the product of Indian and Spanish—we are maturing now; we are coming of age. We are a new nation.

Judy Baca

The Chicano movement stimulated artists as well as political thinkers. Many Mexican American painters drew on the rich tradition of Mexican muralists. After the Revolution, artists such as Diego Rivera, David Siqueiros, and José Clemente Orozco painted huge historical and mythical pictures on the walls of public buildings. Mural painting had been part of the Mexican American tradition earlier, but in the 1970s it took on greater energy and public scope.

One of the pioneers of the mural movement in Los Angeles was Judy Baca. Born in that city in 1946, Baca graduated from California State University. She went on to study art education and traveled to Mexico to learn mural techniques at Cuernavaca. After her return, she started the first mural project in Los Angeles in 1974. Two years later, she and others founded the Social and Public Art Resource Center at nearby Venice. She served as its director during the 1970s.

The project for which Judy Baca is best known is *The Great Wall of Los Angeles,* painted over five summers. The mural is half a mile long, featuring many scenes of Los Angeles's history. To complete the project, Baca employed 450 youths of many cultures and 40 assistant painters as well as other workers as a support staff. Currently, she is working on *World Wall: A Vision of the Future without Fear.*

During the 1920s, Mexican artists such as Diego Rivera created many wall murals that emphasized Mexico's pride in its history and traditions. Their work inspired a similar movement in many Mexican American communities. This is part of a mural created by an artist in Chicago in the 1980s.

The Chicano movement brought many Mexican Americans into politics for the first time. Julia Yslas was born in Mexico in 1910 and moved to the United States with her parents when she was six. Yslas married Emilio Vélez in 1939, but when he died at the age of 46, she had to support their four children. She took a series of secretarial jobs, and finally, in 1965, decided to run for office.

I was fifty-five years old when I entered politics. I was on the South Tucson City Council for twelve years. I went into politics because I love people; I love crowds. And also because it might help my boys—you know—get better jobs. It paid $100 a month, and that helped my kids. I was the first woman on the South Tucson City Council.... I used to go to Washington and fight for my people. I would tell them that there were people here who didn't have running water, electric lights, or indoor plumbing. They were cooking indoors with wood stoves. While I was on the council, we...built two homes for the elderly on East 29th Street—El Señorial and the Bernie Sedley Home. My name is on one of the plaques. We paved some of the streets and put in lights. I have always enjoyed helping the poor....

I remember attending a meeting in Washington, D.C., while I was on the council. It was a black caucus. I loved the way this black lady spoke: "I want black faces in high places." So I took that for my motto. "I want Mexican faces in high places." This is the advice I gave my children: "I want you to make good in school, because I am the poorest in the family, and that is the inheritance I want to leave you—a good education." They believed it and followed it.... It was a hard life, but we made it.

The Chicano movement gave many young Mexican Americans a sense of optimism about their future. They felt that through exercising political power, they could achieve the success that other ethnic groups have gained. In the late 1970s, two 17-year-old Chicanos responded to an interviewer's question about the importance of voting.

Isabel:

We Chicanos have to vote for people who will help us. I mean, if there are candidates who will help us, we will be given more opportunities to get ahead in life.

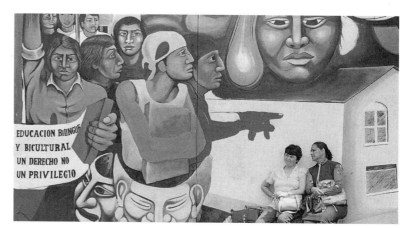

I mean, look at other ethnic groups, they were able to do it that way—like the Irish and the Italians and the Jews. I could name a lot more. They were all part of groups that were discriminated against and they used politics to get a better life.

Jorge:

Hey, Chicanos have to use the ballot to get some power so they can make some changes in those institutions that make it hard for us to get better paying jobs. If we vote correctly, we should be able to get the same high-paying jobs, you know, the ones where there is a lot of status, that the Irish or Jews or other ethnic groups have. After all, they got their jobs through playing politics, by getting all their people to vote a certain way.

After the burst of enthusiasm for La Raza Unida in the 1970s, the movement declined. Some members became disillusioned. A former city councilwoman in a small southern Texan town noted:

When we started the whole thing there were more people involved. About 1,300 people voted for us. Then the gringos cut us down by taking people to court, things like that. Now people are basically afraid of going out to vote. Even the ones who are legal, who would have no problem of being penalized, don't want to go because they don't want any trouble for themselves. The gringos have the power, the money, everything. It is hard to go against them. Like the gringas who have Mexicanas as maids, tell them who to vote for. They even come here to pick up their maids to go and vote, and they give them the sample ballot. So the Mexicanas, of course, vote for the gringos. It is very difficult to make people aware of political problems because of the gringo power over the financial lives of the Mexicanos.

A Mexican American leader in southern Texas described his attitude today.

We, and I, have come full circle. I grew up in South Texas accepting racism. I went off to college and joined the *movimiento* and rallied against racism. Now I find myself working hard to improve this community for both Anglos and Mexicanos. There isn't any other way. We all have to learn to live together.

A Chicano leader who called himself Aguirre commented on the meaning of the Chicano movement:

In our search for identity we are searching and...perhaps creating the concept of *Aztlán*.... The Aztlán people had a civilization that is still with us in a modified form through Mexican-Sapanish influences. Aztlán lives in any land where a Chicano lives: in his mind and heart and in the land he walks. The emblem used extensively in Chicano circles is a black Aztec eagle on a red background.

A Mexican Independence Day parade in the Mexican American barrio of East Los Angeles in 1963. John Moreno, a state assemblyman, is seated at left in the car.

A garment contractor in Los Angeles explained why Mexican immigration is hard to control:

I know a little clothing factory where they rounded up 21 illegals one night and sent them back to Mexico. The next week 19 of them were back working in the same place.... Of course they're going to keep coming. Without the Mexicans, our factory would shut down. If they don't find a way to get across, we'll help them.

A group of young Mexican Americans in Chicago in the 1980s.

NEWCOMERS

Immigrants from Mexico continue to arrive in the United States today, in perhaps greater numbers than ever before. In many ways, their experiences are similar to those who came earlier, but they face what is sometimes a bewildering new world.

Tito is a teenager in Los Angeles. He arrived in the United States when he was four and recalled what growing up here was like.

The first two years after we emigrated, we lived with another Mexican family. When I was about six, we moved to this house, and now eight years later, even I've seen the neighborhood change. There used to be a big orange tree across the street on an open field. I love climbing trees. I'd climb that one, then lay back and eat my fill of fruit. The leaves gave me a nice, cool breeze. The shade, it was great....

One day when I was twelve I heard shouts and trucks and equipment. I looked out the window. The orange tree was gone. Now [drug] dealers and gang members hang out in front of the apartment building that was built. I concentrate on life inside my house with my family, but sometimes it makes me feel like a caged bird.

Vidal Olivares was born in Jalisco, Mexico, but came to the United States in 1976. Today, he works as a truck driver in Los Angeles. He described his life to an interviewer.

For those of us who are raised in a pueblo in Mexico, here we find a life that is really nice, beautiful. One becomes accustomed to the life and doesn't want to return. Now that I have my family, I am planning to stay, if they let us. If someday they send me and my family back, we will return [to the United States], because I like living here very much.

I think it is good that my children will speak both languages. My oldest daughter is four years old and goes to preschool. She can speak various words, count to thirty, and say the days of the week, all in English. At times she surprises me. I get home from work and she says, "Poppy, you have a happy face." She can sing in English too....

Well, to my way of thinking, I don't know if it's good or bad, but what I want now is that my daughters grow up here and learn English. That is, if they don't throw us out. When they are older we'll take them to know Mexico. If one day we can get our papers, then it's all the better. That's how I think.

Julia Garcia de Morales's husband, Faustino, left her and their five children behind when he left Mexico in 1975. Six years later, Faustino wrote that Julia could join him in Chicago. Julia took care of her own children until they were old enough to be left alone. Then she found work caring for the children of a doctor.

I...took care of the doctor's children from the time they were babies. The oldest one is now nine years old. They call me mama because I have raised them since they were little.... But last year the mother of these children decided to stay home.

So I went out and found another job. I began working in McCormick. I worked there six hours daily. It was like a cafeteria.... I would just sweep and collect garbage, mop the floor when people would spill sodas or coffee. I didn't have to wash dishes because everything was disposable. But then in December I had to tell them I was going to Mexico [for the holidays]. I told them I would be available again when I returned, but no, they hired another *señora*. They called me again, but I told them no, because I didn't want to take work away from the *señora*. She has needs too. We're all here trying to live.

I found this other little job in February. I can't explain my work to my friends in San Juan [Mexico]. We make collars for dogs and cats, all kinds of things for animals. No one would understand it. In San Juan you can't catch the cats to put a collar on. It's not something people there would buy.

Right now we're filling orders for Christmas. We fill these socks with things for dogs, little bones, squeaky toys, and things. I know it's kind of crazy, but you can't believe how many orders we have. It's easy, we have chairs and everything.... There isn't any pressure except you have to be very careful that the boss doesn't see you talking, because he will always bawl you out.

Armando Guerra, an editor of the Spanish-language newspaper La Opinión, *with his son Alex in 1968.*

Federico F. Peña

President Bill Clinton's selection of Federico F. Peña as U.S. secretary of transportation in 1993 sent reporters scurrying for information. Few people outside Colorado had heard of Peña. Yet Clinton—who was called the "comeback kid" because of his recovery from political setbacks—admired Peña's ability to overcome adversity.

Public service is a long-standing tradition in Peña's family. Growing up in Laredo, Texas, in the 1950s, he learned that one of his Tejano great-grandfathers had been mayor of Laredo during the Civil War. A grandfather served as a city alderman for more than 25 years.

However, after Peña graduated from the University of Texas law school in 1972, he moved to Denver to open a law firm with his older brother. Peña worked for the local chapter of the Mexican American Legal Defense and Education Fund and joined the Chicano Education Project, which promoted increased funding for schools in Chicano neighborhoods.

Peña's decision to run for mayor of Denver in 1983 seemed unlikely to succeed. One poll showed him with only 3 percent support. But Peña's tireless campaign, with its slogan, "Denver: Imagine a Great City," helped him to win a surprising victory.

To stay in touch with constituents, Peña published his home phone number—until complaints poured in at all hours of the night. Unfortunately, his first term as mayor came during a slowdown in the oil, mining, and high-tech industries that were important to Denver's economy. When Peña ran for reelection, the polls again showed he could not win. Once more, he proved them wrong.

Peña made transportation his special area of expertise. He won support for construction of a huge new airport and rebuilt the city's neglected roads and bridges. In 1991, the first election year when his popularity was high, he decided not to run for a third term. He remained active in public affairs, helping to draw up a 20-year transportation plan for Colorado.

Peña's appointment as secretary of transportation was no surprise to those who knew his record of leadership. As one friend said, "Once he locks into a project, he's like a bulldog.... The tougher it gets, the more he likes it."

MAKING IT

There are countless success stories among today's Mexican Americans. Pancho Lopez, the son of a migrant worker, described his.

Growing up in Mexico, I loved to hear my father's stories about the United States. We couldn't afford for me to go to school after I was twelve. My father was working six months a year in the San Fernando Valley as a farm worker, and he convinced me to come to the States to work because the pay was better. I got a job on a horse farm and I really enjoyed working with the animals. After a while I realized I didn't want to go back to Mexico with my father. I felt like I'd have a much better future working with the horses here in the States. I didn't know English at that time, but once I decided to stay and learn about the horse business I picked up English fairly quickly. It's not so hard when you are young.

I've been in the horse show business for twenty-four years now. I started out with a lot of desire to learn and have acquired a lot of knowledge over the years. The only way to get good at anything is to stick with it. I've only had four employers in all the years I've been in this business, and each time I changed jobs, it's been a step up. You can't build anything up for yourself if you just flit around. My job now involves a lot of responsibilities. I'm in charge of arranging the transportation for dozens of horses all over the world. I hire the other workers and have to know about every part of the business. I always keep my eye out to help other Mexicans have a chance at a good job in the United States. It's one of the bonuses of being a manager in this business. I get to give people a chance to make a good life for themselves.

Like members of other immigrant groups, Mexican Americans felt a sense of pride when they were able to buy their own homes.

Manuel Montemayor, formerly a temporary "green-card" worker, tells how he obtained U.S. citizenship.

As a boy I grew up only seventy miles from the Rio Grande. The first time I crossed the border, I was fourteen years old. I could only get a card for seventy-two hours. I loved it the first time I was in the United States. As a boy, I loved the ice cream and candies.

When I was eighteen, I got a job across the border and stayed in Texas for three years. I met my wife at my job, we went back to Mexico to get married, but then we came back to work for the same family for seven years. I had no green card, but my oldest daughter was born in the United States, so she was the first U.S. citizen in our family. When we went to visit our family in Mexico, we were not allowed back in the United States, so we had to stay in Mexico for three years. Our second child was born there. Finally, my employers helped me file my papers and get back into the country.

I had a job in Chicago after that, but I did not like the big city. I like Texas better because it is near Mexico. We raised our daughters speaking Spanish, and they learned good English in school. My children are Mexican-American. We all love to visit Mexico, but we like to live here. My house will be all paid for in three years. I have a truck and a car. There are much more things to buy here. There is much more opportunity to do different things. There is more work and my children can get a good education. I want my daughters to go to college.

My wife went to school to learn English. I learned English by working. Our oldest daughter only spoke Spanish when she started school, but she learned English quickly and taught her two little sisters a lot before they started school.

We enjoy our life here. I drive the whole family to Disney World, Six Flags in Dallas, and Astro World in Houston. I get to drive my boss all over the United States—to Cleveland and New York. I like where I live, here in San Antonio, the best. Mostly we eat Mexican food in our house. My girls all go to a Catholic school. Here is better because if you want education for your children, you can get it. My older sister and I are the only ones to come to the United States out of ten children. I have had the same employer for twelve years. When I have the time, I like to go to the Gulf Coast to fish, but most of my free time I take care of my house, painting and fixing it up. I enjoy it. I like working with my hands and keeping busy.

Getting my American citizenship was a happy day for me. It is much easier for me to cross the border to see my parents. I can have more influence in the community where I live. My wife and I study the history of the United States and take a test. We have to know how many stars and stripes are on the flag. You have to want to make the effort. I vote every year in every election. I like to watch the news every night.

Lee Treviño

On June 18, 1968, Lee Treviño entered the clubhouse of the Oak Hill Country Club in Rochester, New York. He had just won the U.S. Open Golf Championship. A reporter asked him what he was going to do with the prize money. "I'm going to buy the Alamo and give it back to Mexico," he replied. "I'm the happiest Mexican in the world right now." It was a triumphant moment for a man who had been caddying and shining shoes at a golf course for $30 a week just two years earlier.

Lee Buck Treviño was born in Dallas, Texas, on December 1, 1939, the son of Mexican immigrants. When Lee was five, his uncle gave him an old, rusty golf club and several used golf balls. Before long, he sneaked into a nearby country club course to play a few holes.

After Lee finished the eighth grade, he had to leave school to go to work. Luckily, he found a job at a driving range; after hours, he would hit 300 to 400 balls a day, honing his skills. After a stint in the marines, Treviño moved to El Paso, where he worked at a country club. Sometimes, he challenged other golfers that he could beat them using only a makeshift club with a soda-bottle head. (He usually won.) Treviño turned pro in 1966.

Golf made Treviño a rich man. Although he did not buy the Alamo, he remembered his own background of poverty and was generous in establishing scholarships and supporting causes for the less fortunate. When he retired in 1985, he had the third highest earnings of anyone who had ever played the game.

Vilma Martínez

Vilma Socorro Martínez was born in San Antonio, Texas, in 1943, the daughter of a construction worker. An important influence on Vilma's early life was her grandmother, a very strong woman who taught her to read and write Spanish before she attended school. It gave her confidence when she entered a school where the classes were taught in English. She knew she could learn to read and write in this language as well.

Although Vilma was a good student, when she was ready for high school, her counselor recommended that she go to a vocational school, only because Vilma was Mexican American. But she insisted on enrolling in an academic high school. The same process was repeated when she decided to apply to college. Nevertheless, she enetered the University of Texas and completed her studies in two and a half years. After graduation in 1964, she received a scholarship to study law at Columbia University.

Vilma went into law with the desire to help people. Her first job was at the National Association for the Advancement of Colored People (NAACP) Legal Defense Fund. She brought lawsuits against employers accused of race and sex discrimination. Meanwhile, she heard of a new Mexican American organization being formed to secure civil rights for her people. It was the beginning of the organization that would become the Mexican American Legal Defense and Education Fund (MALDEF).

In 1973, Martínez became president of MALDEF. Though the organization was understaffed, Martínez directed an aggressive strategy. She sued to win the right of children who did not speak English to receive bilingual education in the public schools. Realizing that Mexican Americans did not have adequate political representation, she brought a lawsuit against the state legislature of Texas. She claimed that the state's "at large" system, in which representatives were elected statewide rather than in geographical voting districts, deprived minorities of representation. She won the case. In her years as president of MALDEF, she also raised funds and increased the staff, making it a more formidable organization.

Born in Tijuana, Mexico, Susana Hurtarte became a successful business entrepreneur in the United States. After attending San Diego State College in California, she moved to New Orleans, where she found work as a Spanish-language interpreter. She left that job to raise a family but found that her language skills kept her in demand.

On several occasions clients and business associates of my former employer asked me to substitute for their Spanish-speaking personnel who were either ill or on vacation. I soon recognized the need for an employment agency...that would provide substitutional bilingual personnel to industry, and opened Multi-Lingual Temporary Service with a partner in 1980. Our...staff included persons who could produce translations in several languages and clerical workers assigned exclusively to oil-related companies....

I joined the Latin American Chamber of Commerce. At one of their meetings the district director of the U.S. Small Business Administration [SBA] asked the audience for suggestions to increase Hispanic participation in their programs. I began relating negative experiences shared by Hispanics in the past. At his request I sent him a report outlining ways to improve relations with Hispanic entrepreneurs through uncomplicated methods of counseling.... SBA provided a grant to implement the program through the establishment of the Americas Group. An associate and I were assigned to head the agency.... In 1986 we sponsored a business development seminar and trade fair that later became an annual event known as Latino. The first conference consisted of workshops on marketing, banking, international trade, career improvement, and a trade fair. Over 150 persons attended the meeting. Because the event has grown in scope and attendance, the third annual conference, held in 1988, included the Louisiana Hispanic Chamber of Commerce as one of its sponsors.

A Mexican American father and daughter in the 1990s share the experience of learning about new technology.

Joaquin Avila, Jr., whose parents emigrated from Mexico to Los Angeles, was accepted by Harvard Law School, where he became interested in civil rights law. After graduation, he joined the staff of the Mexican American Legal Defense and Education Fund (MALDEF). After spending several years in California and Texas working for the organization, Avila became president of MALDEF in 1982. Now married and a father, he moved to Fremont, California.

It's not by choice that we live in a white neighborhood. We decided to live in a neighborhood where there were good schools. The children identify themselves as Mexicans. But they don't connect to the culture by language. They know they're different from their classmates, and the oldest one knows that there's discrimination against Hispanics.

He says the discrimination is very subtle. He'll say, "Well, you don't get invited to certain parties. You can hang around with some friends and you can't hang around with other people." It's very subtle, it's not like people telling you they don't want to hang around with you because you're Mexican, but he does feel it....

He wants to go back to Compton [in California, where Avila's grandparents live], because that's where a lot of his relatives are. He feels less isolated when he goes to Compton and spends time with his cousins. In Compton, the areas he spends time in are the barrios. He likes Compton. Even though they're having all these killings [because of gang violence], he feels more at ease. I want him to grow up in a neighborhood where people aren't shooting at each other, where one can feel relatively safe. Whether that's a black or a white or a brown community, to me personally, it doesn't make any difference....

I was the first one out of all my cousins in the United States that went on to college. I would like my children to go on to college. I would like them to go on and have a happy life, a good family life, a good working life, a good work environment. I would like them never to forget that they're Mexicans, never to forget where their parents and grandparents came from.

A Mexican American resident of Chicago in the 1980s, proud of owning his own butcher shop.

113

Edward James Olmos

In the 1980s no Mexican American actor was more popular than Edward James Olmos. He appeared on the stage, in movies, and on a top television show. Audiences knew him as Martin Castillo, the drug-busting police lieutenant on TV's *Miami Vice;* as Jaime Escalante, the brilliant teacher who transformed a class of potential dropouts in a Los Angeles barrio into high-scoring math scholars in the film *Stand and Deliver;* and as the Mexican American hero of the play *The Ballad of Gregorio Cortez.*

Growing up in the Mexican American barrio of East Los Angeles, Olmos taught himself to play the piano and formed his own rock-and-roll band, Eddie James and the Pacific Ocean, which performed in nightclubs along the Sunset Strip.

At the same time, Olmos attended college and took acting lessons. In the 1970s, he started to get bit parts on television and in local theaters. But it was his role in Luis Valdez's play *Zoot Suit* that gave Olmos his career breakthrough. Playing El Pachuco, the narrator and commentator on the action, he won the praise of critics. He would later repeat the role on Broadway and in the motion picture.

Olmos felt especially proud of his work in *The Ballad of Gregorio Cortez,* a play that appeared on public television. It was performed completely in Spanish and told the true story of a man who had been falsely accused of a crime and hunted down by the Texas Rangers in the 19th century.

Olmos uses his celebrity to do charity work and to visit Mexican communities, urging young people to pursue their goals. In 1992, he produced and starred in *American Me,* a film about violence in the barrio of Los Angeles.

His enormous talent has kept him in demand among producers, but Olmos selects his roles carefully, looking for projects that will project a positive image of Mexican Americans and educate other Americans about their rich heritage.

CELEBRATING THE HERITAGE

Tito, a Mexican teenager in Los Angeles, talks about the importance of his heritage.

If you are Mexican, inside you, around you, you feel the Indian and the Spanish cultures. I am both these cultures. On my mom's side, her dad was an Indian, an Aztec. His parents were Aztecs, too. Before my mother was born, her dad fought in the Revolution for a better life and better education for the poor people. Her mom was half Spanish and half Indian. Her mother's mother was from Spain. Her mother's father was also an Indian. Her grandfather went and stole her and they got married....

In many ways my dad is American now. He doesn't speak Spanish that well, but he understands everything. And even though my mother doesn't speak English, she understands it. When my father and my mother talk, she speaks to him in Spanish and he answers her in English. I talk to my dad in English and my mother in Spanish. I'm used to it; it's what I've always known. What matters is that I know what I'm about and so do the important people in my life.

I don't understand kids who are angry at their parents and their culture. You are what you are. You can't do anything about it. If somebody says, "Oh, Mexicans can't do this," I want to prove them wrong. Mexicans have gone through so much hardship. I wish this country realized more what we give back.

Joaquin Avila, Sr., left his home in Chihuahua, Mexico, in 1946 and moved to Los Angeles. Four decades later, his son Joaquin, Jr., a graduate of Yale University, was president of the Mexican American Legal Defense and Education Fund [MALDEF]. Joaquin Avila, Sr., told an interviewer how he raised his son to preserve his cultural roots.

I used to sit down with Joaquin, Jr., with the paper *La Opinión,* and he learned to read when he was about five or six, from reading *La Opinión.* The same with Jaime [his younger son]. In my house, we had to speak only Spanish. I knew they would learn English in school, but here it was only Spanish. And I think it paid off; it was good for them. To speak another language is to be more educated and to have more culture. It's beautiful to know four or five languages even, but I wouldn't let them forget their mother tongue....

I think you have to talk to your children about their roots, so they'll have an idea where they came from, what country their parents come from. You should tell them stories, anecdotes, and they begin to understand and to like history. Let's talk about Junior, for example. Mexican history is very beautiful to my son, to the extent that he gave his children the names of Mexican leaders. His oldest son is named for Zapata. His younger son is named Salvador Tiachka. Tiachka is an Indian name; it means the man who gives advice, a wise man. Everyone calls him Salvador, but my wife calls him Tiachka.

On November 1 and 2, Mexicans and Mexican Americans celebrate the Days of the Dead. It is a time to remember deceased family members. Zarela Martínez, who now lives in New York City and runs a fashionable restaurant, recalled a memorable celebration on an island in Michoacán called Janitzo.

People were selling toys in the shape of laughing skulls; candies made like skeletons; chicle gum (cousin to our chewing gum) in all colors; black, yellow, or white candles; candleholders of shiny black pottery; *copal* (incense) burners. Streams of cheerful family parties pushed past us in the dark and confusion as if en route to a midnight picnic—which isn't too far off the mark.

Janitzo is a rocky island, and the hilly cemetery might have been hard to navigate had it not been so full of lights. It was a carnival lit by thousands of candles. Hundreds of brilliant orange-yellow shapes—high arches or crisscrossed squares—glowed in the darkness above every gravesite. They were *cempasúchil,* yellow marigolds strung on ropes or string, sometimes spelling out names. I had never seen or dreamed of such a fantasy of *cempasúchil.* Huddled in shawls and *rebozos* [mufflers] against the chilly evening, the people settled down to sing and drink and chat on the ground by the graves.... The strings of flowers were arches framing a passage for the souls of their friends or relatives to return to their earthly resting place on a kind of social visit. This is the whole purpose of the celebration....

Beyond the favorite foods of the deceased, certain things are traditional for the Days of the Dead that date back to pre-Columbian times, including a sweet pumpkin dessert called *calabaza en tacha* and some forms of *tamal....* But the most famous specialty of the day...is of European origin. In one of those typical Indian-Spanish intermarriages that have shaped our culture, the native peoples came to celebrate the Days of the Dead with a rich, sweet yeast bread on the model of the altar breads that are special feast-day offerings everywhere in Europe from Spain to Sweden. The Mexican imagination put a new spin of fantasy on the idea by shaping the loaves into different images. The famous *pan de muerto* ("bread of death") comes in the shape of human figures, alligators, lizards, and other animals—but most often skulls and crossbones or teardrops and crosses gaily decorated with colored sugar crystals.

A Mexican American family in Los Angeles wears makeup for the Day of the Dead festivities.

Zarela Martínez, a Mexican immigrant, owns a fashionable restaurant (named Zarela) in New York City. She enjoys decorating her tables with traditional Mexican folk art that she has collected. Here, a Day of the Dead skeleton is served with a plate of tamales.

José Limón

Sometime around 1930, José Limón sat in a darkened theater in New York City watching a ballet. As he watched the male dancer on the stage, Limón was stunned by the strength as well as the beauty of the performance. "What I saw simply changed my life. I saw the dance as a vision of ineffable power." Limón would go on to become one of the most important figures in classical dance in the 20th century.

José Limón was born in 1908 in the western town of Culiacán, Mexico. As a boy, he enjoyed the art, music, and folk dances of his native country. When he was seven, José and his family immigrated to the United States to avoid the ravages of the Mexican Revolution. The Limóns settled in Los Angeles.

When he was 20, Limón went to New York City to study painting. His teachers complained that his work was old-fashioned. Depressed, Limón left the art school, unsure of what he would do with his life. It was in this low period that he saw the life-changing ballet performance.

Afterward, Limón made a pledge to himself. He would achieve two goals—to become a great dancer and a great choreographer, or designer of dances. Immediately, he plunged into dance lessons, at an age when it is usually too late to begin them. In the 1930s he started getting jobs in Broadway musicals. The money was good, but it did not fulfill his desire to do something different in modern dance. "Modern dance is not a popular art," he would later explain. "It is not for us to advertise automobiles, rugs, vacuum cleaners or hair dyes."

Limón started to develop his own vision, designing works that attacked war, poverty, and racial injustice. Limón also incorporated Mexican themes into his art. Audiences were enthralled and by the end of the 1940s, he was a regular performer in the New York City dance season and took his art on tour throughout the world.

In the last decades of his life, Limón devoted himself to teaching dance. Presidents John F. Kennedy and Lyndon Johnson both asked him to perform for White House galas. At his death in 1973, he was recognized as one of the great masters of modern dance.

Holidays like Christmas, Easter, and many saints' days are often associated with special kinds of food. Mike Acosta of San Antonio recalled one special Christmas-season dish—buñuelos.

Buñuelos are in the shape of a tortilla but they are made from a batter that is deep fried with lots of sugar and cinnamon. They are something that we usually eat during Christmas and during the holidays. They can also be shaped as a butterfly.

Mike Acosta's daughter Bernice added:

At Christmas time, I always remember the night before Christmas Eve all the granddaughters would get together at my grandmother's house and make tamales. We took the masa or the dough or the corn husks and wrapped meat in them.

Around Easter time, my grandmother Acosta would make *gorditas,* which are really fat, deep fried corn tortillas with a meat filling, lettuce and tomato, all the fixings. *Gordita* means "fat woman," and I guess if you eat enough of them that's what it will make you!

Also, on Sunday mornings sometimes my dad would bring home *pan de guste,* which is sweet bread, and *barbacoa,* which is...roasted cow head (laughter)—it's true, sounds kinda gross, but really good if you're raised on it. And tripe, which also sounds awful because it's cow's stomach, fried in a big metal skillet at Easter, Christmas Eve, and after weddings, when all the family and friends would get together to celebrate. Tripe is also used in *menudo,* a stew that's made with hominy and a chili pepper seasoning.

During the time of Lent—the 40 days before Easter that starts on Ash Wednesday—it's a Catholic tradition not to eat meat on Fridays. My mother never cooked meat on Fridays during Lent. Also, in any Mexican restaurant in San Antonio, they always had Lenten specials, which was a big meal that excluded meat but had fish or just vegetable entrees.

In the early 20th century, Mexican American women in San Antonio sold food in Alamo Plaza. These women, called "chili queens," probably introduced to the United States the spicy stew known as chili.

Musicians at a fiesta in New Mexico.

A Charro Days celebration in Brownsville, Texas. Charro is a nickname for a vaquero when he dresses in an elaborately decorated outfit. During Charro Days, people celebrate their Mexican heritage by wearing traditional clothing, like the dresses worn by these girls.

A boy in traditional costume prepares to take a swing at this elaborate piñata at a fiesta on Olvera Street in Los Angeles.

The third and fourth generations of the Acosta family in the United States. From left, Bernice Acosta-Cole, Mike Acosta, Michael Acosta, Jr., Belinda Acosta, and Olga Acosta, Mike's wife.

Refugia Acosta Alonso as a girl, around 1925.

THE ACOSTA FAMILY

San Antonio is the largest city in the United States with a Mexican American majority—about 60 percent of its nearly 1 million people. Founded by Spanish soldiers 118 years before the Battle of the Alamo was fought there, San Antonio has been a magnet for Mexican immigrants throughout the 20th century.

The Acosta family arrived in 1920. Today, on the city's west side stands a little shop with the sign "Acosta Music Company." Three generations of Acostas have carried on this business, which is now operated by Mike Acosta. His aunt, Refugia Acosta Alonso, was five years old in 1916 when the family crossed the Rio Grande from Mexico, but she still remembers the trip.

Because of the fighting, people tried to get out. People went hungry, even if they had money, because the government changed the money [issued new currency, making the old bills valueless]. Then my grandfather said, "I am going to get out of here," because at that time he belonged to the Estado Major, men of Pancho Villa. And his oldest son was only 15, but the government was taking young men [into the Mexican army].

We took a train, and we came to Nuevo Laredo [Mexico] where the revolution was already going on. And they had burned half of the bridge so that people couldn't come across. A lot of people had no place to go, and they were all over the city. They lived in shacks or tents that had been donated by the U.S. army. We lived there until the fighting stopped and then came across to Laredo [Texas], where we had a nice room. My daddy [Guadelupe] used to play in a band and also to make guitars. My brother and I used to help him. In Nuevo Laredo he made a communion rail and a pulpit for the church. I haven't been there for a long time, but it's still there. In fact the priest of the church was the godfather for [her brother] Miguel, who was born in Nuevo Laredo.

In 1920 we moved to San Antonio, because my grandmother, my mother's mother, was already here. Maybe you heard about the chili queens that used to live here in San Antonio way back in the old days. Over at the farmer's market, in the square. Well at that time, at about 4:30 in the morning old ladies used to sell chili there. My grandmother was one of them. She would sell menudo, chili con carne, and tamales. The Southern Pacific Railroad ran through here, and a lot of Mexican people worked on the tracks. They would get up at five or six o'clock and want something to eat. So that's where grandmother made her money. And by six or seven o'clock she had sold everything.

Q: *Why did your father decide to start the music store?*

Refugia:

Sí, he was a very independent man. He didn't like to work for anybody. He had made the bajo sexto in Mexico with my grandfather, and by the time he came here he was skilled. He was a perfectionist. He had a special place where he would get the wood. He was very careful, picked out just the right kind.

At that time everything was done by hand, no machines. A typical day, he would start in the morning by eight, seven-thirty. Sometimes people would want a guitar quickly, and he would work hard till nine o'clock at night.

But when he decided to start his business, it was a sacrifice for all of us. Nowadays, you have Christmas, kids have a tree, get all kinds of presents. The next day they throw it away, because they get too much. On our Christmas, my daddy would say to my mother, "What do the children need?" And she would say this one needs shoes, that one needs a coat. And we were satisfied with that, because we needed it.

Outside of that we used to have a lot of food. Tamales, buñuelos, champurals. You make champurals with masa [corn meal], add some water to it, then strain it and put in some cinnamon and sugar. Then you toast it until it becomes thick like Cream of Wheat. It was a hot meal and that was good to eat in the winter. So we used to have a lot of that, because my grandmother would make a lot of good things.

Three of Guadelupe Acosta's sons joined the family business, and it became the largest musical instrument company in San Antonio during the 1930s and 1940s. Miguel Acosta, the youngest son, became widely known for the quality of the guitars he made, and he often used them to perform in public. Miguel's oldest son, Mike, keeps the Acosta Music Company going today.

Q: *When did you join the family business?*

Mike:

I started in the shop in 1956, when I was in high school. Because of the music that was popular at that time, there was a big demand for guitars. My father had pretty much closed down, but then I started things again. But I just run the shop in the evenings now, repairing guitars, violins, other instruments. The guitars you see hanging on the wall are rebuilt and repaired ones—not handmade.

The business thrived in the 1930s and 1940s. Back then, they had a print shop where they printed their own songbooks as well. All of us kids used to do the binding and folding at night, after school. The books were sold all over the country, wherever there was a Mexican community.

At that time my father could sell a handmade guitar for about $145. But you couldn't make a living doing that now. With machine-made guitars available, you couldn't charge enough for the work of a handmade one.

A musical group organized in the 1920s by Guadalupe Acosta (standing at right). The young man with the guitar is Guadalupe's son Miguel, who is the father of Mike Acosta.

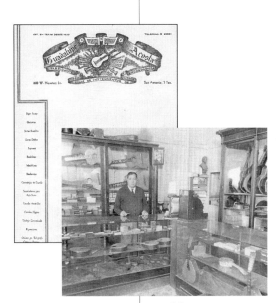

Guadalupe Acosta in the family music store in 1931. The store's elaborate stationery listed, in Spanish, the many kinds of instruments and supplies it sold.

In 1937, Guadalupe and Miguel Acosta posed for a series of photographs showing the steps in making a handcrafted guitar. At top, Guadalupe shapes the sides of the guitar. In the middle, Guadalupe makes guitar necks. At bottom, Miguel finishes the bond between neck and body.

Q: This picture shows a double-neck guitar that your father made in 1947. How did he get the idea for it?

Mike:

Well, after he came back from serving in World War II, in 1945, he used to play with some local bands. And whenever they didn't have a bass player, or couldn't afford one, he had to play a bajo sexto, a 12-string bass guitar. So as a guitar-maker, he came up with the idea of a double-neck guitar. It was a combination of a regular six-string and a bajo sexto. Unfortunately, he never did patent his idea. And in 1947, he used it with an electric amplifier as well. It was practical and made sense, because he didn't have to carry two guitars. Later the big guitar manufacturers came out with the same thing, but he was the first to develop it.

Q: What was it like to grow up in San Antonio? Did you ever experience prejudice?

Mike:

I don't know if you can call it prejudice, but there weren't very many advantages for Mexican Americans in our school. Mostly our school was a vocational school. There weren't many college preparatory courses. They trained you for carpentry, painting, stuff like that.

At one time I worked for a pharmacy and I wanted to save money to enter pharmacy school. I went to the dean's office, and they told me that was a waste of time—that I was supposed to be a carpenter, and I had to take carpentry courses, graduate from high school and get a job to help my parents. There was no encouragement to take college courses.

Q: What about your children? Have they lost a sense of their Mexican Spanish heritage?

Mike:

Yes. I don't think nowadays kids care so much about it. They're more interested in getting ahead, making money, having a better life-style. My kids didn't have it as hard as we did.

Mike's children—Bernice, Belinda, and Michael, Jr.—the fourth generation of Acostas in the United States—describe their own experiences.

Bernice:

Well, one thing about San Antonio is that even though it is the tenth-largest city in the nation it still has that small-town feel to it. When I go to visit friends in Houston and Dallas, those cities are kind of overwhelming. And San Antonio has the Spanish influence and culture. All the festivities like fiesta, the block festivals, the architecture, the music—are Hispanic.

Belinda:

Growing up in San Antonio, one thing I remember is the influence of the Catholic community. I spent the last two years of my high school education in public schools and that was the first time I ever met Protestants or Jews. I think the reason is that most of the festivities in San Antonio just surround the Catholic community—parish festivals. The greatest cultural influence in the city is Catholic, and most people are involved in them.

Michael, Jr.:

One thing that I remember about the family is the closeness. Even when I lived in different parts of the country, I found that Hispanic families seem to be close. And even if you're not part of the family, other Hispanics seem to welcome you in and treat you like their own, which I think is a very important part of the culture.

One of the things I remember that I still enjoy are family gatherings. I can't count all the times we've gotten together for Christmas or Easter or someone's birthday. We're talking about 20, 30, or 40 people just sitting down and eating and drinking and celebrating. I think that's what being Hispanic is all about—family. I think another part is the Church, which has been an important part of my life. I've been an altar boy. I was in a Christian music group in high school, attended Catholic school. And I think it's a meeting place that is part of our culture.

Q: Do you think you will carry on these traditions with your own family?

Michael, Jr.:

I think I will. I think it's very important. Today, with all the violence, I think family values are important to instill in our children. I'm getting married in November and I think it's very healthy to be family oriented.

Q: Bernice, when you got married, you kept the name Acosta. Sometimes you hyphenate it with your husband's name, Acosta-Cole. Why is that?

Bernice:

Well, marrying a non-Hispanic as well as a non-Catholic, I feel that it's important that my children are aware of their background. They will be exposed to both cultures, both traditions, and I want them to enjoy and appreciate the Hispanic culture as well as their other culture.

In this 1951 photo, Guadalupe's daughter Sylvia holds one of the double-neck guitars that her father and brother (at right) invented.

A 1990s family reunion. Miguel Acosta, Sr., and his wife, Octavia, stand in the center.

MEXICAN AMERICAN TIMELINE

1519
Spanish conquistadors conquer the Aztec Empire of Mexico.

1598
Juan de Oñate begins Spanish colonization of what is now the southwestern United States.

September 16, 1810
With *grito de Dolores*, Father Miguel Hidalgo starts Mexican independence movement.

1821
Mexico wins independence from Spain.

1836
Texans revolt against Mexican rule.

1845
Texas becomes part of the United States.

1846–48
U.S.–Mexican War

1848
Treaty of Guadalupe Hidalgo

1850
California becomes a state.

1854
Gadsden Purchase

May 5, 1862 (*Cinco de Mayo*)
Mexicans defeat a French invading army.

1894
Alianza Hispano Americana founded in Tucson, Arizona.

1911–20
Mexican Revolution, during which many refugees flee to the United States.

1924
U.S. Border Patrol is established.

1929
League of United Latin American Citizens (LULAC) founded in Corpus Christi, Texas.

1930s
Repatriation program sends many Mexican American immigrants back to Mexico.

1935
Dennis Chávez of New Mexico becomes first U.S. senator of Mexican descent.

1941–45
More than 300,000 Mexican Americans serve in U.S. armed forces in World War II.

1942–47, 1951–64
Bracero program recruits Mexicans to work in the United States.

1954–58
U.S. government carries out Operation Wetback to return illegal workers to Mexico.

1962
César Chávez starts National Farm Workers Association, forerunner of United Farm Workers union.

1965–70
La Huelga, the strike against grape growers around Delano, California.

1965
Bilingual Education Act provides funds to encourage teaching in the Spanish language.

1965
U.S. Immigration and Nationality Act sets a limit of 120,000 immigrants yearly from Western Hemisphere.

1968
Founding of Mexican American Legal Defense and Education Fund (MALDEF).

1970
Patrick Flores becomes first Mexican American bishop of the Catholic church.

1986
Immigration Reform and Control Act offers amnesty to certain groups of illegal aliens, including agricultural laborers.

1993
North American Free Trade Agreement (NAFTA) reduces trade barriers between Mexico, United States, and Canada.

FURTHER READING

General Accounts of Mexican American History

Alford, Harold J. *The Proud Peoples: The Heritage and Culture of Spanish-Speaking Peoples in the United States.* New York: David McKay, 1972.

Four Hundred Fifty Years of Chicano History in Pictures. Albuquerque, N. Mex.: Chicano Communications Center, 1976.

McWilliams, Carey. *North from Mexico.* Westport, Conn.: Greenwood Press, 1986.

Meier, Matt S., and Feliciano Rivera. *The Chicanos: A History of Mexican Americans.* New York: Hill & Wang, 1972.

Samora, Julian, and Patricia Vandel Simon. *A History of the Mexican-American People.* Notre Dame, Ind.: University of Notre Dame Press, 1977.

Shorris, Earl. *Latinos: A Biography of the People.* New York: Norton, 1992.

Specific Aspects of Mexican American History

Chávez, John R. *The Lost Land: The Chicano Image of the Southwest.* Albuquerque: University of New Mexico Press, 1984.

Galarza, Ernesto, Herman Gallegos, and Julian Samora. *Mexican-Americans in the Southwest.* Santa Barbara, Calif.: McNally & Loftin, 1970.

Langley, Lester D. *MexAmerica: Two Countries, One Future.* New York: Crown, 1988.

Morin, Raul. *Among the Valiant: Mexican-Americans in World War II and Korea.* Alhambra, Calif.: Borden, 1966.

Moquin, Wayne, with Charles Van Doren. *A Documentary History of the Mexican-Americans.* New York: Praeger, 1971.

Muñoz, Carlos, Jr. *Youth, Identity, Power: The Chicano Movement.* New York: Verso, 1989.

Sánchez, George J. *Becoming Mexican American: Ethnicity, Culture and Identity in Chicano Los Angeles, 1900–1945.* New York: Oxford University Press, 1993.

First-Person Accounts of Mexican American Life

Acosta, Oscar Zeta. *The Autobiography of a Brown Buffalo.* San Francisco: Straight Arrow Books, 1972.

Beeson, Margaret, Marjorie Adams, and Rosalie King. *Memories for Tomorrow.* Detroit: Blaine Ethridge Books, 1980.

Bode, Janet. *New Kids on the Block.* New York: Franklin Watts, 1989.

Chegin, Rita Kasch. *Survivors: Women of the Southwest.* Las Cruces, N. Mex.: Yucca Tree Press, 1991.

Davis, Marilyn P. *Mexican Voices/ American Dreams.* New York: Henry Holt, 1990.

Galarza, Ernesto. *Barrio Boy.* Notre Dame, Ind.: University of Notre Dame Press, 1971.

Gamio, Manuel. *The Mexican Immigrant.* New York: Arno Press, 1969.

Martin, Patricia Preciado. *Images and Conversations: Mexican Americans Recall a Southwestern Past.* Tucson: University of Arizona Press, 1983.

Martin, Patricia Preciado. *Songs My Mother Sang to Me: An Oral History of Mexican-American Women.* Tucson: University of Arizona Press, 1992.

Owsley, Beatrice Rodriguez. *The Hispanic-American Entrepreneur: An Oral History of the American Dream.* New York: Twayne, 1992.

Poggie, John R., Jr., ed. *Between Two Cultures: The Life of an American-Mexican.* Tucson: University of Arizona Press, 1973.

Rodriguez, Richard. *Hunger of Memory: An Autobiography.* New York: Bantam, 1982.

Siems, Larry, trans. and ed. *Between the Lines.* Hopewell, N.J.: Ecco Press, 1992.

Poems, Stories, Novels, and Plays by Mexican Americans

Cisneros, Sandra. *The House on Mango Street.* Houston: Arte Publico Press, 1985.

Cisneros, Sandra. *Woman Hollering Creek.* New York: Random House, 1991.

Harth, Dorothy E., and Lewis M. Baldwin. *Voices of Aztlán: Chicano Literature of Today.* New York: New American Library, 1974.

Ortega, Philip D., ed. *We Are Chicanos: An Anthology of Mexican-American Literature.* New York: Washington Square Press, 1973.

Valdez, Luis, and Stan Steiner, eds. *Aztlán, An Anthology of Mexican American Literature.* New York: Knopf, 1972.

Villaseñor, Victor. *Macho!* New York: Bantam, 1971.

Villaseñor, Victor. *Rain of Gold.* Houston: Arte Publico Press, 1991.

TEXT CREDITS

Main Text

p. 12, top: John R. Chavez, *The Lost Land* (Albuquerque: University of New Mexico Press, 1984), 12.

p. 12, bottom: Luis Valdez and Stan Steiner, eds., *Aztlán, An Anthology of Mexican American Literature* (New York: Knopf, 1972), 44-45.

p. 13: David J. Weber, ed., *Foreigners in Their Native Land* (Albuquerque: University of New Mexico Press, 1973), 47-49.

p. 14: Cathy Luchetti, *Home on the Range: A Culinary History of the American West* (New York: Villard, 1993), 211-12.

p. 15: Reprinted from *Across the Tracks: Mexican-Americans in a Texas City* by Arthur J. Rubel, copyright © 1966. By permission of the author and the University of Texas Press.

p. 16, top: Wayne Moquin with Charles Van Doren, *A Documentary History of the Mexican-Americans* (New York: Praeger, 1971), 203-4.

p. 16, bottom: Weber, *Foreigners in their Native Land,* 169-73.

p. 17: Moquin and Van Doren, *A Documentary History,* 106-7.

p. 18, top: Louis Adamic, *From Many Lands* (New York: Harper & Brothers, 1940), 51.

p. 18, bottom: Fabiola Cabeza de Baca, *We Fed Them Cactus* (Albuquerque: University of New Mexico Press, 1954), 6-8.

p. 19 and p. 20: Rita Kasch Chegin, *Survivors: Women of the Southwest* (Las Cruces, N. Mex.: Yucca Tree Press, 1991), 143-49.

p. 21: Patricia Preciado Martin, *Images and Conversations* (Tucson: University of Arizona Press, 1983), 55-56.

p. 26 and p. 27, top: From *Barrio Boy* by Ernesto Galarza. © 1971 by University of Notre Dame Press. Reprinted by permission of the publisher, 51-54.

p. 27, bottom: Margaret Beeson, Marjorie Adams, and Rosalie King, *Memories for Tomorrow* (Detroit: Blaine Ethridge Books, 1983), 47-48.

p. 28: Reprinted from *Latinos, A Biography of the People,* by Earl Shorris, by permission of the author and W. W. Norton & Company, Inc. Copyright © 1992 by Earl Shorris.

p. 29: Marilyn P. Davis, *Mexican Voices/American Dreams* (New York: Henry Holt, 1990), 303-4.

p. 30 and p. 31: Beeson, Adams, and King, *Memories for Tomorrow,* 50-52.

p. 32, top and bottom: Beeson, Adams, and King, *Memories for Tomorrow,* 53-54.

p. 33: Beeson, Adams, and King, *Memories for Tomorrow,* 2.

p. 34, top: Oscar Lewis, *Pedro Martínez: A Mexican Peasant and his Family* (New York: Random House, 1964), 84.

p. 34, middle and bottom: From *Barrio Boy* by Ernesto Galarza. © 1971 by University of Notre Dame Press. Reprinted by permission of the publisher, 111-12.

p. 35 and p. 36: Beeson, Adams, and King, *Memories for Tomorrow,* 62-66.

p. 36, bottom: Henry P. Anderson, *The Bracero Program in California* (Berkeley: University of California Press, 1961), 82.

p. 37: Al Santoli, *New Americans: An Oral History* (New York: Viking, 1988), 268.

p. 42, top: Davis, *Mexican Voices/American Dreams,* 132-33.

p. 42, bottom: Reprinted from *American Mosaic: The Immigrant Experience in the Words of Those Who Lived It,* by Joan Morrison and Charlotte Fox Zabusky, by permission of the University of Pittsburgh Press. © 1980, 1993 by Joan Morrison and Charlotte Fox Zabusky, 347-49.

p. 43: Davis, *Mexican Voices/American Dreams,* 114-15.

p. 44, top: Reprinted from *American Mosaic: The Immigrant Experience in the Words of Those Who Lived It,* by Joan Morrison and Charlotte Fox Zabusky, by permission of the University of Pittsburgh Press. © 1980, 1993 by Joan Morrison and Charlotte Fox Zabusky, 121.

p. 44, bottom: John J. Poggie, Jr., ed., *Between Two Cultures* (Tucson: University of Arizona Press, 1973), 59-62.

p. 46, top: Martin, *Images and Conversations,* 6-7.

p. 46, bottom: Charles J. Bustamante and Patricia L. Bustamante, *The Mexican-American and the United States* (Mountain View, Calif.: Patty-Lar Publications, 1969), 34.

p. 47, first: Reprinted from *Latinos, A Biography of the People,* by Earl Shorris, by permission of the author and W. W. Norton & Company, Inc. Copyright © 1992 by Earl Shorris.

p. 47, second: *Washington Post,* June 6, 1993, A-1.

p. 47, third: *Washington Post,* June 6, 1993, A-27.

p. 47, fourth: Janet Bode, *New Kids on the Block* (New York: Franklin Watts, 1989), 71.

p, 48: Santoli, *New Americans: An Oral History,* 285-86.

p. 49, top: Davis, *Mexican Voices/American Dreams,* 262-63.

p. 49, bottom: Davis, *Mexican Voices/American Dreams,* 315-21.

p. 50: Larry Siems, trans. and ed., *Between the Lines* (Hopewell, N.J.: The Ecco Press, 1992) 9-11.

p. 51: Siems, *Between the Lines,* 43-45.

p. 54, top: Thomas E. Sheridan, *Los Tucsonenses* (Tucson: University of Arizona Press, 1986), 75.

p. 54, middle: Beeson, Adams, and King, *Memories for Tomorrow,* 66-67.

p. 54, bottom: Poggie, *Between Two Cultures,* 22-23.

p. 55: Poggie, *Between Two Cultures,* 72-73.

p. 56, top: Douglas E. Foley et al., *From Peones to Politicos: Class and Ethnicity in a South Texas Town, 1900<n>1987* (Austin: University of Texas Press, 1988), 14.

p. 56, bottom: Foley, *From Peones to Politicos,* 54.

p. 57, top: Foley, *From Peones to Politicos,* 118-19.

p. 57, bottom: Foley, *From Peones to Politicos,* 59-60.

p. 58: Patricia Preciado Martin, *Songs My Mother Sang to Me* (Tucson: University of Arizona Press, 1992), 212-14.

p. 59, top: Beeson, Adams, and King, *Memories for Tomorrow,* 65-66.

p. 59, bottom: Emory S. Bogardus, *The Mexican in the United States* (Los Angeles: University of Southern California Press, 1934) 21-22.

p. 60: Reprinted from *American Mosaic: The Immigrant Experience in the Words of Those Who Lived It,* by Joan Morrison and Charlotte Fox Zabusky, by permission of the University of Pittsburgh Press. © 1980, 1993 by Joan Morrison and Charlotte Fox Zabusky, 122-23.

p. 61, top: Foley, *From Peones to Politicos,* 38.

p. 61, bottom: Ronald Takaki, *A Different Mirror* (Boston: Little, Brown, 1993) 323.

p. 62: Foley, *From Peones to Politicos,* 89-90.

p. 63, top: From *Barrio Boy* by Ernesto Galarza. © 1971 by University of Notre Dame Press. Reprinted by permission of the publisher, 262-63.

p. 63, middle: Raul Morin, *Among the Valiant* (Alhambra, Calif.: Borden, 1966) 21-22.

p. 63, bottom: Anderson, *The Bracero Program in California,* 8.

p. 64, top: Reprinted from *American Mosaic: The Immigrant Experience in the Words of Those Who Lived It,* by Joan Morrison and Charlotte Fox Zabusky, by permission of the University of Pittsburgh Press. © 1980, 1993 by Joan Morrison and Charlotte Fox Zabusky, 350.

p. 64, bottom: Elizabeth Loza Newby, *A Migrant with Hope* (Nashville, Tenn.: Broadman Press, 1977), 17-18, 27, 29-30.

p. 65: Erasmo Gamboa, ed., *Voces Hispanas: Hispanic Voices of Idaho* (Boise: Idaho Commission on Hispanic Affairs and Idaho Humanities Council, 1992), 20-21.

p. 66: Eliot Wiggonton, ed., *"I Wish I Could Give My Son a Wild Raccoon"* (Garden City, N.Y.: Doubleday Anchor Press, 1976), 83.

p. 67, top: Beeson, Adams, and King, *Memories for Tomorrow,* 68.

p. 67, bottom: Reprinted from *American Mosaic: The Immigrant Experience in the Words of Those Who Lived It,* by Joan Morrison and Charlotte Fox Zabusky, by permission of the University of Pittsburgh Press. © 1980, 1993 by Joan Morrison and Charlotte Fox Zabusky, 350-351.

p. 68, top: Sheridan, *Los Tucsonenses,* 72-73.

p. 68, bottom: Susan Garver and Paula McGuire, *Coming to North America from Mexico, Cuba, and Puerto Rico* (New York: Delacorte, 1981), 29.

p. 69, top: Clarke Newlon, *Famous Mexican-Americans* (New York: Dodd, Mead, 1972), 54.

p. 69, middle: Morin, *Among the Valiant,* 20.

p. 69, bottom: Foley, *From Peones to Politicos,* 51.

p. 74: Valdez and Steiner, *Aztlán,* 54-57.

p. 75: Gamboa, *Voces Hispanas,* 12-13.

p. 76: From *Barrio Boy* by Ernesto Galarza. © 1971 by University of Notre Dame Press. Reprinted by permission of the publisher, 200-206.

p. 77: Oscar Zeta Acosta, *The Autobiography of a Brown Buffalo* (San Francisco: Straight Arrow Books, 1972), 71-76.

p. 78, top: Davis, *Mexican Voices/American Dreams,* 330-31.

p. 78, bottom: Martin, *Images and Conversations,* 65.

p. 79, top: Foley, *From Peones to Politicos,* 123.

p. 79, bottom: Foley, *From Peones to Politicos,* 55.

p. 80, top: Victor Villaseñor, *Rain of Gold* (Houston: Arte Publico Press, 1991), 558.

p. 80, bottom: Chegin, *Survivors: Women of the Southwest,* 27.

p. 81: Reprinted from *Across the Tracks: Mexican-Americans in a Texas City* by Arthur J. Rubel, copyright © 1966. By permission of the author and the University of Texas Press.

p. 82, top: Reprinted from *Across the Tracks: Mexican-Americans in a Texas City* by Arthur J. Rubel, copyright © 1966. By permission of the author and the University of Texas Press.

p. 82, bottom: Davis, *Mexican Voices/ American Dreams,* 370-71.

p. 84, top: Adamic, *From Many Lands,* 243-44.

p. 84, bottom: Foley, *From Peones to Politicos,* 108.

p. 85, top: Manuel P. Servin, *An Awakened Minority* (Beverly Hills, Calif.: Glencoe Press, 1970), 284-85.

p. 85, middle: Reprinted from *Latinos, A Biography of the People,* by Earl Shorris, by permission of the author and W. W. Norton & Company, Inc. Copyright © 1992 by Earl Shorris.

p. 85, bottom: Gamboa, *Voces Hispanas,* 17.

p. 86: Martin, *Songs My Mother Sang to Me,* 62-66.

p. 87, top: Martin, *Songs My Mother Sang to Me,* 133.

p. 87, bottom: Martin, *Songs My Mother Sang to Me,* 72-73.

p. 88, top: Richard Rodriguez, *Hunger of Memory* (New York: Bantam, 1982), 81, 85-86.

p. 88, bottom: Rodriguez, *Hunger of Memory,* 92.

p. 89: David F. Gomez, *Somos Chicanos: Strangers in Our Own Land,* 16.

p. 90: Martin, *Songs My Mother Sang to Me,* 210-12.

p. 91, top: Reprinted from *Latinos, A Biography of the People,* by Earl Shorris, by permission of the author and W. W. Norton & Company, Inc. Copyright © 1992 by Earl Shorris.

p. 91, bottom: From *Barrio Boy* by Ernesto Galarza. © 1971 by University of Notre Dame Press. Reprinted by permission of the publisher, 206.

p. 92, top and bottom: Chegin, *Survivors,* 30-31.

p. 93, top: Martin, *Songs My Mother Sang to Me,* 70.

p. 93, bottom: Morin, *Among the Valiant,* 24, 83, 87-88.

p. 94: Reprinted from *Latinos, A Biography of the People,* by Earl Shorris, by permission of the author and W. W. Norton & Company, Inc. Copyright © 1992 by Earl Shorris.

p. 100: Valdez and Steiner, *Aztlán,* 203-4.

p. 101, top: Servin, *An Awakened Minority,* 287-88.

p. 101, bottom: Ellen Cantarow, *Moving the Mountain: Women Working for Social Change* (Old Westbury, N.Y.: The Feminist Press and McGraw Hill, 1980) 129-37.

p. 102: Studs Terkel, *Working* (New York: Pantheon, 1974), 37-38.

p. 103, top: César Chávez, Sister Mary Prudence, and Louis Valdez, "Huelga! Tales of the Delano Revolution," *Ramparts 5* (July 1966): 42.

p. 103, bottom: Garver and McGuire, *Coming to North America from Mexico, Cuba, and Puerto Rico,* 69.

p. 104, top: Lynn P. Dunn, *Chicanos: A Study Guide* (San Francisco: R&E Research Associates, 1975), 108.

p. 104, bottom: "Are You a Chicano?" *DQU Newsletter,* October 17, 1972.

p. 105, top: Rodolfo "Corky" Gonzales, *I am Joaquín* (New York: Bantam, 1972).

p. 105, bottom: Jack D. Forbes, ed., *Aztecas Del Norte: The Chicanos of Aztlán* (Greenwich, Conn.: Fawcett, 1973), 287.

p. 106, top: Martin, *Songs My Mother Sang to Me,* 36-37.

p. 106, bottom: Martin Sanchez Jankowski, *City Bound* (Albuquerque: University of New Mexico Press, 1986), 126, 128.

p. 107, top: Foley, *From Peones to Politicos,* 211-12.

p. 107, bottom: Foley, *From Peones to Politicos,* 285-86.

p. 108: Bode, *New Kids on the Block,* 68-70.

p. 109, top: Davis, *Mexican Voices/ American Dreams,* 208-9.

p. 109, bottom: Davis, *Mexican Voices/ American Dreams,* 219-20.

p. 110: Mary Motley Kalergis, *Home of the Brave: Contemporary American Immigrants* (New York: Dutton, 1989), unpaged.

p. 111: Kalergis, *Home of the Brave.*

p. 112: Beatrice Rodriguez Owsley, *The Hispanic American Entrepreneur* (New York: Twayne, 1992), 35-37.

p. 113: Reprinted from *Latinos, A Biography of the People,* by Earl Shorris, by permission of the author and W. W. Norton & Company, Inc. Copyright © 1992 by Earl Shorris.

p. 114, top: Bode, *New Kids on the Block,* 72-73.

p. 114, bottom: Reprinted from *Latinos, A Biography of the People,* by Earl Shorris, by permission of the author and W. W. Norton & Company, Inc. Copyright © 1992 by Earl Shorris.

p. 115: Zarela Martínez, *Food From My Heart* (New York: Macmillan, 1992), 204-6.

p. 116: personal interview conducted by Kevin Cole, 1994.

Sidebars

p. 14: Ronald Takaki, *A Different Mirror* (Boston: Little, Brown, 1993), 170.

p. 15: Cathy Luchetti, *Home on the Range: A Culinary History of the American West* (New York: Villard, 1993), 213.

p. 17: Roger Daniels, *Coming to America: A History of Immigration and Ethnicity in American Life* (New York: Harper, 1990), 314.

p. 19: David J. Weber, ed., *Foreigners in their Native Land* (Albuquerque: University of New Mexico Press, 1973), 235-36.

p. 28: Margaret Beeson, Marjorie Adams, and Rosalie King, *Memories for Tomorrow* (Detroit: Blaine Ethridge Books, 1983), 49-50.

p. 29: Beeson, Adams, and King, *Memories for Tomorrow,* 49.

p. 31: Beeson, Adams, and King, *Memories for Tomorrow,* 46.

p. 34: Merle E. Simmons, *The Mexican Corrido as a Source for Interpretive Study of Modern Mexico* (Bloomington: Indiana University Press, 1957), 348.

p. 36: Joshua Freeman et al., *Who Built America?* (New York: Pantheon, 1992), 243.

p. 37: Manuel Gamio, *The Mexican Immigrant: His Life Story* (Chicago: University of Chicago Press), 4, 32.

p. 46: Luis Alberto Urrea, *Across the Wire* (Garden City, N.Y.: Doubleday Anchor Books, 1993), 20.

p. 48: Beatrice Rodriguez Owsley, *The Hispanic American Entrepreneur* (New York: Twayne, 1992), 35.

p. 55: Takaki, *A Different Mirror,* 319.

p. 58: William V. Wells, "The Quicksilver Mines of New Almaden," *Harper's Magazine,* June 1863, 30.

p. 61: Takaki, *A Different Mirror,* 332.

p. 63: Reprinted from *Latinos, A Biography of the People,* by Earl Shorris, by permission of the author and W. W. Norton & Company, Inc. Copyright © 1992 by Earl Shorris.

p. 64: Carey McWilliams, *Ill Fares the Land* (Boston: Little, Brown, 1942) 267.

p. 65: Albert Camarillo, *Chicanos in a Changing Society* (Cambridge: Harvard University Press, 1979), 172.

p. 68: *New York Times,* June 20, 1920.

p. 69: Carey McWilliams, *North From Mexico* (New York: Greenwood Press, 1968), 249.

p. 75: Reprinted from *Across the Tracks: Mexican-Americans in a Texas City* by Arthur J. Rubel, copyright © 1966. By permission of the author and the University of Texas Press.

p. 81: Reprinted from *Across the Tracks: Mexican-Americans in a Texas City* by Arthur J. Rubel, copyright © 1966. By permission of the author and the University of Texas Press.

p. 85: Rita Kasch Chegin, *Survivors: Women of the Southwest* (Las Cruces, N.Mex.: Yucca Tree Press, 1991), 32-33.

p. 87: Wayne Moquin with Charles Van Doren, *A Documentary History of the Mexican-Americans* (New York: Praeger, 1971), 286-87.

p. 91: Joan W. Moore with Alfredo Cuéllar, *Mexican Americans* (Englewood Cliffs, N.J.: Prentice-Hall, 1970), 144.

p. 93: Louis Adamic, *From Many Lands* (New York: Harper & Brothers, 1940), 257-58.

p. 106: Lynn P. Dunn, *Chicanos: A Study Guide* (San Francisco: R&E Research Associates, 1975), 2.

p. 107: Dunn, *Chicanos: A Study Guide,* 2.

p. 108: Thomas Kessner and Betty Boyd Caroli, *Today's Immigrants: Their Stories* (New York: Oxford University Press, 1982), 73-74.

PICTURE CREDITS

Courtesy the Acosta family: 118-19, 120 middle and bottom, 121; Archives of Labor and Urban Affairs, Wayne State University: 101 bottom, 103; Arizona Historical Society Library: 20 bottom (#2138), 21 (#29039), 58 (Mexican Heritage Project, #64323), 70 (Mexican Heritage Project, #64444), 78 (Mexican Heritage Project, #64313), 84 bottom (Mexican Heritage Project, #69472), 91 (#47870), 95 bottom (Mexican Heritage Project, #63527); Arte Publico Press: 49; Bancroft Library: 8, 12; Center for Migration Studies/Fr. Ezio Marchetto, C.S.: 42 top and bottom, 43, 45 top and bottom, 46; K. Chapman, courtesy Museum of New Mexico, #57564: 13; courtesy Rita Kasch Chegin: 20 top, 85; courtesy Billy Deal: 111; Gerald Dean, U.S. Department of Housing and Urban Development: 6-7; The Denver Public Library, Western History Department: 59; Southwest Collection, El Paso Public Library: 25, 31, 34, 35 top and bottom, 36, 37, 40; El Pueblo de Los Angeles Historic Monument: 74 top, 117 bottom; Thomas Hoobler: 115 bottom; courtesy Houston Metropolitan Research Center, Houston Public Library: 83 top right, 104; Reproduced by permission of the Huntington Library, San Marino, California: 10, 14, 16 top, 18 bottom, 54 top, 66 top; Impact Visuals: 24 (Philip Decker), 38 (David Maung), 44 (David Maung), 83 top left (Amy Zuckerman); Institute of Texan Cultures, San Antonio, Tex.: 16 bottom, 54 bottom, 56 top and bottom, 57 top and bottom, 73, 116, 117 top left, 120 top; Library of Congress: 15, 47, 48, 50, 51, 52, 55, 60 top and bottom, 61, 63, 64, 74 bottom, 76 bottom, 79, 82 top, 92 top, 117 top right; Library of the Daughters of the Republic of Texas at the Alamo: 11, 17, 28; Mexican American Legal Defense Fund: 112 top; Minnesota Historical Society: 92 bottom; courtesy Museum of New Mexico #22468, 18 top; National Archives: 62 bottom, 76 top, 84 top; The Billy Rose Theatre Collection, The New York Public Library for the Performing Arts, Astor, Lenox and Tilden Foundations: 102; The Dance Collection, The New York Public Library for the Performing Arts, Astor, Lenox and Tilden Foundations: 116 top; Arthur Howard Noll and A. Philip McMahon, *The Life and Times of Miguel Hidalgo y Costilla* (New York: Russell & Russell, 1910): 27 top; Ohlinger: 114; Antonio Perez: 86, 88, 89 top and middle, 96, 98, 106, 108, 112 bottom, 113 bottom; Santa Barbara Historical Museums: 66 bottom; Security Pacific Collection/Los Angeles Public Library: cover, frontispiece, 5, 62 top, 69, 72, 75, 77 top and bottom, 80, 82 bottom, 83 bottom, 89 bottom, 93, 95 top, 99, 101 top, 107, 109, 110 bottom, 113 top, 115 top; Seaver Center for Western History Research, Natural History Museum of Los Angeles County: 90; Suzanne Seriff, Texas Folklore Society: 41; Social and Public Art Resource Center, Venice, California: 105; Phil Stern, courtesy UCLA Photo Collection: 67; Archives Division—Texas State Library: 33; Underwood Collection, Bettmann Archive: 22, 26 bottom, 27 bottom, 29, 32; United Nations Photo #155564/Pat Goudvis: 30; UPI/Bettmann: 65; courtesy U.S. Department of Transportation: 110 top.

INDEX

References to illustrations are indicated by page numbers in *italics*.

Acosta, Bernice, 116, 120-21
Acosta family, 118-21
Acosta, Mike, 116, 119-21
Acuna, Roberto, 102-3
Adobe, 54
Aguilera, Josefina, 35-36
Alamo, 23
Alianza de los Pueblos Libres, 105
Alianza Hispano Americana, 72, *91, 97*
American Coordinating Council for Political Education (ACCPE), 97
American G.I. Forum, 97
Anaya, Rodolfo A., 99
Anguilar Melantzón, Ricardo, 49-50
Apodaca, Jerry, 99
Art, 13, 98, 105
Assimilation, 71-73, 108-13
Associations, 90-91, 97
Avila, Joaquin, Jr., 113, 114
Avila, Joaquin, Sr., 114-15
Avila, Margarita, 94
Aztecs, 9-10, 12
Aztlán, 9, 12, 98, 107

Baca, Judy, 105
Bandits, 17-18, 19
Barrios, 71, *72,* 74-77, 98, 113
Battle of San Jacinto, 23
Bilingual Education Act (1965), 99
Border Patrol, *38,* 39, 41-45, 49-50
Braceros, 40, *55,* 63-64
Brown Berets, 98
Brownsville, Texas, 39, *117*
Buñuelos, 15, 116

Caballero, Cesar, 48
Cabeza de Baca, Fabiola, 18-19
Californios, 13-18
Carranza, Jacinta, 78-79
Castillo, Ana, 99
Castro, Raul, 99
Celebrations, *29, 83,* 115-16, *117*
Chávez, César, 68, 100-3
Chávez, Dennis, 73
Chicano movement, 98, 104-7
Chicanos, 98, 104-7
Chilton, Elsie Chavez, 80, 85, 92
Cinco de Mayo, 24, 71, 91
Cisneros, Henry G., 6-7
Cisneros, Sandra, 51, 99
Citizenship, U.S., 111. *See also* Immigration
Ciudad Juárez, Mexico, 25, 39, 42, 49

Club Filarmónico, 21
Compadrazgo, 72, 81
Contratista, 62-63
Coronel, Antonio Franco, 16-17
Corridos, 19, 33, 36, 77
Cortés, Hernán, 10
Courtship, 31, 81-82
Cowboys, 12
Coyotes, 41, 42-43

Days of the Dead, 115
Delano, California, strike. See *La Huelga*
Delgado, Socorro, 86-88, 93
Díaz, Porfirio, 24-25, 32
Diego, Juan, 10
Dolores, Mexico, 10, 27
Domestic workers, 55
Durán, Diego, 12

Education. *See* Schools
El Paso, Texas, 10, 11, 20, 25, 39, 42, 48, 49-50, 77
Escalante, Frank, 68

Family values, 27-28, 71-72, 78-82
Farm workers, 25, 53, 54-55, 60-65, 97-98, 100-103
Fernández, Celestino, 49
Fiestas, 14, 90, 99, *117,* 120
First Communion, 88-89
Flores, Patrick, 71
Food, 10, 12, 15, 99, 115, 116, 119. *See also individual foods*
Fraga, Felix, 91
Fuentes, Rubén, 21
Funerals, 87-88

Galarza, Ernesto, 26-27, 34-35, 63, 76-77, 91-92
Gangs, 72
Garcia de Morales, Julia, 109
García, Hector, 97
Garcia, José, 67-68
Garibay, Miguel, 29, 33
Garza, George, 63
Gender roles, 27-29, 71-72, 78-79, 82
Gold rush of 1849, 11, 16-17
Gomez, David F., 46, 89, 98
Gonzales, Ramón, 44-45, 54-55
Gonzales, Rodolfo "Corky," 98, 105
González, Henry B., 97
Great Depression, 40, 63, 73, 92-93
Grito de Dolores, 10, 27, 90
Guerrero, Julio, 47
Gutiérrez, José Angel, 98

Hayworth, Rita, 90
Herbal medicine, 20-21, 31
Hernández Tovar, Jesús Manuel, 42
Hidalgo, Miguel, 10, 27, 90
Hispanos, 11, 72
Holidays, 14-15, 29, 30, 71, 75, 76, 86-87, 90, 91, 99, 115-17, 119
Huerta, Dolores, 85, 100, 101
Huerta, José, 54
Huerta, Victoriano, 25
Hurtarte, Susana, 112

I Am Joaquín (Gonzales), 98, 105
Immigration, 39-41, 46-51
    due to Mexican Revolution, 34-36
    due to poverty, 36-37
    illegal, 42-45
Immigration and Nationality Act (1965), 40

Jalcocotán, Mexico, 26
Juárez, Benito, 23

Ku Klux Klan, 66

La Bamba, 102
La Causa, 100
La Huelga, 97-98, 100-3
Land grant disputes, 16-18
Language, 11, 25, 85, 99
La Opinión, 109, 114
La Raza, 9
La Raza Unida, 98, 104, 107
Laredo, Texas, 11, 39, 110, 118
La Sociedad Benito Juárez, 72
La Sociedad Fraternal Moctezuma, 72
League of United Latin American Citizens (LULAC), 73, 91, 97
Liga Protectora Latina, 72-73
Limón, José, 116
Lincoln County War, 11, 18
Longoria, Félix, 97
Lopez De La Cruz, Jessie, 101-2
López, Ignacio, 97
Lopez, Pancho, 110
López Tijerina, Reies, 105
Los Angeles, California, 10, *55,* 71, 72, *74, 77, 82,* 98, 108, 109, *117*
Los Lobos, 29-30, 99
LULAC. *See* League of United Latin American Citizens

Machismo, 72, 82
Macías, Lupe, 43
Madero, Francisco, 25

MALDEF. *See* Mexican American Legal Defense and Education Fund
Manifest Destiny, 9
Maquiladoras, 48
Martinez Arrayo, Sabino, 28, 31
Martínez, José P., 94
Martínez, Lucia, 36, *59*
Martínez, Pedro, 34
Martínez, Vilma, 112
Martínez, Zarela, 115
Matamoros, Texas, 39
Maximilian, 24
Mendoza, Isabella, 44, 60-61
Mestizos, 9
Mexican American Legal Defense and Education Fund (MALDEF), 99, 110, 112, 113, 114
Mexican-American Political Association (MAPA), 97
Mexican Revolution, 25, 33-36
Mexico, 9-10, 12-13, 23-25, 26-31
Mexico City, *27*
Midwifery, 20, 31
Migra. See Border Patrol
Migrant laborers, 41, 53, 64-65, 97
Military service, 73, 93-94, 97, 98
Mining, 12-13, 23, 53, 58
Missions, religious, 10, 13
Moctezuma II, 10
Montalban, Ricardo, 99
Montemayor, Manuel, 111
Morales, Martha, 69
Moreno, John, 107
Morin, Raul, 63, 69, 93-94
Murals, 98, 105, 106
Murieta, Joaquin, 11, 17-18, 105
Mutualistas, 72, 73

National Chicano Moratorium Day (August 29, 1970), 98
National Farm Workers Association (NFWA), 97
Newby, Elizabeth Loza, 64
New Deal programs, 73
New Mexico, 9, 11, 12-13, 19-20
New Spain, 10, 12
Nieto, Lupe, 66
Nogales, Arizona, 39
Nogales, Mexico, 39
Nuevo Laredo, Mexico, 39, 118

Olivares, Vidal, 109
Olmos, Edward James, 49, 99, 114
Olvera Street (Los Angeles), *74, 117*
Oñate, Juan de, 10
Operation Wetback, 40
Ortega Robert, 85

Peña, Federico F., 110
Pérez, Rita, 65
Perez Rivas, Maria, 30
Pico, José Ramon, 14-15
Plan of Ayala, 25
Political Association of Spanish Speaking Organizations (PASSO), 97
Politics, 97, 98-99, 104-7
Poverty, 36-37, 41
Prejudice, 16-17, 61, 66-69, 84-85, 91, 97, 120

Quetzalcoatl, 9-10
Quinceañeras, 76, 96
Quintero, Ramiro, 78

Railroad workers, 24-25, 53, 54, 59, 118
Ranch workers, 56-57
Refugees, Mexican Revolution, 34-36
Religion, 30, 31, 71, 86-89
Repatriation programs, 40, 46, 73
Reyes Pitts, Rose, 82
Rio Grande, 24, 39, *40*
Rivera, Diego, 105, 106
Rodriguez King, Dolores, 30
Rodriguez, Richard, 88-89
Rodriguez, Robert, 99
Rodriquez, Sergio, 47
Roman Catholicism, 10, 23, 24, 71. *See also* religion
Ronstadt family, 21
Ronstadt, Linda, 21, 99
Rosas, Cesar, 29-30
Roybal, Edward, 97

Sacramento, California, 76-77
Salazar, Rubén, 98, 106
San Antonio, Texas, 10, 11, 61, 110, 116, 118, 120-21
San Diego, California, 10, 39, 48
Santa Anna, Antonio López de, 23
Santos, 13
Sayre, Elina Laos, 21
Schools, 61, 69, 71, 84-85, 99
Silva, Juanita, 31
Silvas Martin, Carlotta, 58, 90
Siqueiros, David, 105
Sleepy Lagoon case, 67
Sonora, Mexico, 11, 16
Soria, Sotero H., 27-28, 32-33, 67
Soto, Gary, 99
Southwest Voter Registration Education Project, 98-99
Spain, 10-11, 12-13
Suarez, Mario, 74-75

Teatro Campesino, 102, 103
Tejanos, 9, 11, 16
Teotihuacán, 9, 10
Tijuana, Mexico, 42-46, 48
Torres, Miguel, 42
Treaty of Guadalupe Hidalgo, 9, 11, 16, 18
Treviño, Lee, 111
Tuscon, Arizona, 21, 74-75
Tyson, Elizabeth Ray, 55

United Farm Workers (UFW), 97, 100-103
Unity Leagues, 97
Urbina, Rosa María, 37
U.S.-Mexican War, 23

Valdez, Luis, 102, 103, 114
Valdez, Lupe, 84
Valens, Ritchie, 102
Vallejo, Guadalupe, 13-14
Vallejo, Mariano, *8,* 18
Vaqueros, 12, 53
Varela, José, *41*
Vásquez, Abel, 85
Vietnam War, 98
Villa, Pancho, 25, 33, 118
Villaseñor, Victor, 49, 80
Virgin of Guadalupe, 10, 13, 27, 71, 87-88, 89
Voting, 98-99, 107

Wainwright, J. M., 93
Waldrip, Sophie Rodriguez, 19-21
Warren, Adelina Otero, 87
Waxman, Al, 69
Weddings, 13, 31
"Wetbacks," 39
Wilson, Woodrow, 33
Wirtz, Willard, 53
Women, 20-21, 56, 78-79, 94-95
World War I, 73, 75
World War II, 40, 73, 93-95

Yslas, Julia, 106

Zapata, Emiliano, 25, 33
Zaragoza, Ignacio, 24
Zarate, Patricia, 50-51
Zarate Salmeron, Jeronimo de, 12-13
Zazueta Huerta, Juanita, 75
Zeta Acosta, Oscar, 77
Zoot Suit riots, 69, 94

ACKNOWLEDGMENTS

We owe a great debt of gratitude to Carolyn Cole of the Los Angeles Public Library; Diane Bruce of the Institute of Texan Cultures; Barbara Bush, photo librarian of the Arizona Historical Society; Antonio Perez; and Diane Zimmerman and Father Ezio Marchetto of the Center for Migration Studies. This book would not have been possible without their contributions.

Our heartfelt thanks to Rita Kasch Chegin, author of *Survivors: Women of the Southwest,* for her generosity in sharing the photographs she collected and allowing us to publish excerpts from her work.

We would also like to express our appreciation for the valuable contributions and assistance we received from The Arte Publico Press; Tara Deal and Nancy Toff, our editors at Oxford University Press; Thomas Featherstone of the Wayne State University Archives of Labor and Urban Affairs; Joe S. Graham of Texas A&I University; Nancy Hadley of the Houston Metropolitan Research Center; Sandy Hise and Bernice Maggio of the Border Patrol Museum; Angela Holtzman of the Harold Washington Library in Chicago; Zarela Martínez; Alice McGrath; Richard Ogar of the Bancroft Library; Arthur S. Olívas of the Museum of New Mexico; Luis Pedroza of the Chicano Resource Center in Los Angeles; Abelardo de la Peña, Jr., of MALDEF; Jean Poole of the El Pueblo de Los Angeles Historic Park; Dr. Marco Portales of Texas A&M University; Tim Ready of the Science Museum of Minnesota; Michael Redmon of the Santa Barbara Historical Society; Holly Reed of the National Archives; Bill Richter of the Center for American History, University of Texas; Joscelin Sherman of the United Farm Workers of America; Dr. Errol Stevens of the Seaver Center for Western History Research; Kathey Swan of the Denver Public Library Western History Department; Martha Utterback of the Daughters of the Republic of Texas Library; Jennifer Watts of the Huntington Library; and Marion Zientek of the *Texas Catholic Herald.*

Finally, we owe special thanks to the Acosta family, who shared their memories and their own Mexican American family album with us. We will always remember the warm welcome we received from Mike Acosta, who showed us through his music store and workshop when we arrived on very short notice. We are especially grateful to Mike's daughter Bernice for providing a copy of her essay on the family history and to her husband, Kevin L. Cole, for conducting the interviews with the other members of the Acosta family.

ABOUT THE AUTHORS

Dorothy and Thomas Hoobler have published more than 50 books for children and young adults, including *Mexican Portraits; Margaret Mead: A Life in Science; Vietnam: Why We Fought; Showa: The Age of Hirohito;* and *Photographing History: The Career of Mathew Brady.* Their works have been honored by the Society for School Librarians International, the Library of Congress, the New York Public Library, the National Council for Social Studies, and *Best Books for Children*, among other organizations and publications. The Hooblers have also written several volumes of historical fiction for children, including *Frontier Diary, The Summer of Dreams,* and *Treasure in the Stream.* Dorothy Hoobler received her master's degree in American history from New York University and worked as a textbook editor before becoming a full-time freelance editor and writer. Thomas Hoobler received his master's degree in education from Xavier University, and he previously worked as a teacher and textbook editor.